Famine and Survival Strategies

Famine and Survival Strategies

A Case Study from Northeast Ethiopia

Dessalegn Rahmato

Nordiska Afrikainstitutet, Uppsala 1991
(The Scandinavian Institute of African Studies)

This book is published with support from
The Swedish International Development Authority (SIDA)

ISBN 91-7106-314-5

Typeset and printed in Sweden by
Bohusläningens Boktryckeri AB, Uddevalla 1991

Contents

Abbreviations 6

Glossary 7

Acknowledgements 9

Section I
INTRODUCTION

1. Objectives of the Study 13
   Community and the ethic of cooperation 18
2. Organization of the Study 35
   Sources for the Study 36
   Technical problems and usage 43

Section II
FAMINE: HIDING BEHIND THE MOUNTAINS

3. Wollo and Ambassel: The Setting 47
4. The Economy of Wollo 57
5. The Peasant Mode of Production 69
   Farming practices 76
   The control of the micro-environment 81
   Consumption, marketing and prices 87
6. Famine in Wollo 99
   The death toll (1984–85) 107

Section III
SURVIVAL: COMMUNITY AND COOPERATION

7. The Community in Distress 117
   Crisis anticipation 118

Magic and divination 125
Crisis management 141
Exhaustion and dispersal 156
8. Survival Strategies 163
Austerity and reduced consumption 165
Divestment and asset disposal 171
Livestock flows during the famine 176
Normal and abnormal behaviour 182
9. Post Famine Recovery 193

Section IV
BEYOND SURVIVAL
10. Neither Feast Nor Famine 211
Disaster designation and early warning 219

References 227
Information from Official Records 227
Primary Sources 227
Secondary Sources 230

Annexes
1. Rainfall Data, Haiq Station, Ambassel 1963–1984
2. Livestock Supply and Prices, Haiq Market
3. Grain Prices, Bistima Market, Ambassel
4. Recipients of Relief Aid in Wollo 1984
5. National and International Relief Agencies, Wollo 1984–85
6. Emergency Food Aid to Ethiopia 1984 to 1986

MAPS
1. Wollo: Zones and Administrative Divisions 48
2. Ambassel: Livestock Flows During Famine 67

Abbreviations

AACM	Australian Agricultural Consulting and Management
AMC	Agricultural Marketing Corporation
CSO	Central Statistical Office
EC	Ethiopian Calendar

EHRS	Ethiopian Highlands Reclamation Study
EMS	Ethiopian Meteorological Service
ENI	Ethiopian Nutrition Institute
EPID	Extension and Project Implementation Department (of the Ministry of Agriculture)
ESP	Ethiopian Serategnoch Party
FAO	Food and Agricultural Organization
GC	Gregorian Calendar
Ha.	Hectare (equals 2.47 acres)
HASIDA	Handicraft and Small-Scale Industries Development Agency
ILCA	International Livestock Centre for Africa (Addis Ababa)
ILO	International Labour Office (Organization)
LMB	Livestock and Meat Board (Ethiopia)
MOA	Ministry of Agriculture
NEEPR	Northeast Ethiopia Planning Region
NEERNDRC	Northeast Ethiopia Region Natural Disaster Relief Committee (Dessie, Wollo)
NGO	Non-Governmental Organization
NRDC-CPSC	The name of the Central Planning Agency
PA	Peasant Association
RRC	Relief and Rehabilitation Commission
SCF	Save the Children Fund (UK)
WFP	World Food Programme
WPA	Woreda Peasant Association

Glossary

Abbay	Local name for the Blue Nile river.
Awraja, Woreda	The country has a three-tier administrative division. The lowest administrative unit is the *woreda;* several woredas make an *awraja*, and several awrajas make a *province*.
Ensat (Enset)	A long-maturing plant which provides the staple food of the population in south-central Ethiopia.
Kebbelae, got	see Chapter 1.
Quintal	A unit of measurement equal to 100 kg. Ten quintals equal one ton.

Red Cross	The Ethiopian Red Cross Society.
Teff	The staple food of the population of northern Ethiopia.
Woreda	see Awraja.
Calendar	There is a seven to eight year difference between the Ethiopian Calendar (EC) and the Gregorian Calendar (GC). We started our field work in Wollo in Tiqimt 1979 (EC), October 1986 (GC).
Currency	At the official rate of exchange, US$1.00 equals 2.07 (Ethiopian) *Birr* (1986).

Acknowledgements

An earlier draft of this work was distributed in the form of a report, and there I expressed my thanks to all the NGOs which provided me financial and logistical support during the research. I wish to thank here the Social Science Research Council (New York) for its grant which enabled me to finalize the study and prepare the manuscript for publication.

My thanks also go to the many peasants, peasants leaders, and extension agents in Wollo who gave us a good deal of their time during our field-work. I have also greatly benefitted from the support and encouragement of friends and colleagues at the Institute of Development Research (Addis Ababa University), and at the African Studies Center (Boston University) where the final phase of the study was completed.

Finally, I would like to thank the Scandinavian Institute of African Studies for their cooperation in publishing the manuscript.

The views expressed are solely my own and do not reflect those of funding agencies or research institutions.

Dessalegn Rahmato

I INTRODUCTION

1. Objectives of the Study

When we first visited the province of Wollo in northeast Ethiopia in October 1986 to start the field-work for this study, what we saw was not exactly what we had been led to expect. The countryside appeared verdant and full of life, the harvest on many peasant plots was in good condition, and the people, young and old, bore no visible marks of having recently endured a traumatic experience. And yet, only a year and half ago famine had swept through most of Wollo, and death, privation and disease had played havoc with rural society in the province.

The people of Wollo bore the brunt of the 1984–1985 Ethiopian famine which may be considered the worst tragedy that rural Ethiopia has *ever* experienced. The scale of human suffering in the province was such that this alone would have been sufficient to place the country high on the list of the world's most distressed nations in those two fateful years. The casual visitor to Wollo in the last quarter of 1986, however, would hardly suspect that such a terrible disaster had ravaged the area only a short while ago.

But the crisis was not really over, only its surface evidence had temporarily been covered up. The deep wounds inflicted by hunger and disease take a long time to heal, and we could see that the healing process had barely got under way. Paradoxically, and as the past experience of Wollo shows, it is in the years of recovery that the seeds of famine are actually sown.

A severe food crisis immediately followed by an apparent recovery, itself succeeded by another full-scale crisis: such has been the modern history of rural Wollo. The tragedy of the past fuels the tragedy of the future, and the spectre of famine refuses to be exorcised. It is not that the farming population here (and in this study we are concerned with the sedentary agriculturalists—the pastoralists of lowland Wollo will have to be treated in a separate work) is unfamiliar with the fine arts of crop husbandry, nor is it unwilling or unable to labour sufficiently hard. Indeed, we shall try to show later in this study that the indigenous agricultural science of the rural cultivators here—and the same is true of their counterparts in northeast Ethiopia—is fairly well-advanced, although it has not often been given the credit it deserves. The Wollo peasantry is diligent, frugal and highly skilled, and yet this same peasantry has been the victim of all the major famines that have occurred in the country in the last one hundred years.

Wollo has suffered six serious famines since the beginning of this century. Since the 1950s famine has hit the northeast—in which we include Wollo and Tigrai provinces, and the eastern and northern portions of Gondar and Shoa provinces, respectively—almost once every ten years. In the period between 1975 and 1983 there were numerous localized food shortages in one part of Wollo or another, and since 1979 "normal" harvest years and years of death have followed each other in close succession. In brief, Wollo may be described as a land of hunger and privation.

The most recent famine that struck Wollo and the northeast did not start in 1984, it was only discovered by the international media in that year. Documents from the Relief and Rehabilitation Commission (RRC) show clearly enough that famine was building up force and momentum all through the latter half of the 1970s. In 1980, only two of the twelve awrajas of the province—Borena and Wore-Illu in the southwest—were relatively free of food shortages, and by the end of 1982 RRC was reporting that over one million peasants in the province, i.e. about one-third of the rural population, were facing starvation. In December 1984, the same agency reported that more than 75 per cent of the Wollo peasantry were unable to feed themselves and in danger of imminent death. Of the 7.7 million rural Ethiopians said to be seriously suffering from starvation in 1984, 2.6 million, or more than one-third, were Wollo peasants.

Famine has returned to the northeast after an absence of less than two years. As of December 1987, over two million peasants in Tigrai and Wollo were said to be in distress and in need of emergency assistance. In the early part of the year, the northeast and Eritrea were threatened by a serious locust invasion, said to be the worst outbreak since 1958, and while the threat soon passed without much damage it heightened peasant insecurity. The outlook for 1988, according to RRC, was quite gloomy: more than 5.2 million peasants in the country were expected to starve if relief supplies did not reach them in time. The worst affected areas were Eritrea, Tigrai and northern Wollo, but peasants in North Shoa, northeastern Gondar, and parts of Harrarghe provinces were also suffering from drought and food shortages. In many areas in the northern provinces the 1987/88 harvest failed completely, and peasants were faced with starvation on a large scale. The death toll in the worst hit areas had risen sharply since October 1987, but relief supplies had been sent to the areas, and this may have averted a large-scale tragedy.

Wollo may be a land of famine but let no one say it is a land of sorrows. In our travels through many parts of the province we encountered few signs of deep grief, shock or bitterness having to do with the tragedy of the recent past. Many of the peasants we talked to discussed the famine and events connected with it without remorse, and with a good deal of humour. This is not to suggest that the people were unmindful of the terrible loss, human and material, they had so recently suffered. On the contrary, to a people accustomed to living with imminent danger, the living are more important than the dead, and the future of more concern than the past. To us this cheerful attitude, this refusal to brood over what had happened, was a reflection of the strength of character of the rural population, and part of their survival strategies.

The material for this study was collected in Wollo, and the work is almost exclusively about this province. But Wollo is part of the northeast, and often enough what happens here has repercussions in the region as a whole.

The purpose of the study is not to recount the story of Wollo under famine, nor to assess in full the impact of the disaster on the economy and the society. The objective is a limited one: it is to examine the survival strategies of a peasantry faced with famine, and particularly at a time when external assistance is either nonexistant or of very limited reach.

What do peasants do in the face of severe food crisis and ecological stress, and how do they manage to survive on their own? This is the main question that we shall try to answer in the course of this work. However, the subject is a broad one, and leads to a whole series of other related questions and enquiries. How far ahead in time are peasants conscious of an approaching disaster? What are traditional techniques of "early warning" and disaster preparedness, and how effective are they? At what point do the indigenous survival techniques break down or prove inadequate, giving rise to mass migration or mass death? These and similar issues will be discussed with the 1984 and 1985 Wollo famine serving as case material.

Coping with crisis is an important aspect of peasant life. Look closely at the routine activities of the average peasant in any normal season and you will discover that some of what he does is aimed at insuring his family against death and hardship. The decision to grow a particular crop mix in a particular season, to acquire or dispense with this or that farm animal, to sell one kind of produce and buy another... Each of these decisions will involve at least two important considerations: firstly, a purely economic one, i.e. whether or not the peasant will obtain the best possible returns on his investment or labour; and secondly, a "survival" consideration, i.e. whether or not the peasant household will be exposed to privation should an unexpected situation arise shortly after the decision has been made. Crisis anticipation forms a central part of peasant agronomy, and rare are rural communities which do not have elements of "disaster preparedness" built into their farming systems and exchange relations.

A distinction should of course be made between what may be called *anticipatory* survival strategies and *crisis* survival strategies, the former being those adopted during periods of normalcy, and the latter in times of stress. It should be pointed out however that the difference between the two is one of degree, and often the second type of strategy is an extension of the first but magnified or distorted by emergency conditions and the accelerated course of events.

It is frequently noted, for instance, that migration is one form of peasant response to the threat of starvation, but migration in search of employment goes on all the time, although the practice is on a much smaller scale of magnitude in the years when the harvest is good than in those when it is not. It should also be noted that it is the effectiveness of anticipatory strategies that will determine the success of crisis strategies. While both forms of coping will be examined in this work, greater weight will be given to crisis management techniques in periods of serious food shortages.

Peasant response to disaster is not actually a new subject, but it has not been given the importance it deserves in the literature on Ethiopian famines. The general literature on famine response is also, in our opinion, incomplete and unsatisfactory for the following reasons: (a) it frequently singles out one or a few aspects of the subject, and by treating these in isolation, fails to bring out a fuller picture of the problem. (b) It often suggests that coping is an individual effort, and either neglects or seriously underestimates the important role played by the community and the given system of community relations. The view that the peasant household, and very often the household head, is sovereign, and the latter's decision is made independently is quite prevalent. We shall return to this point later. (c) It does not pay enough attention to indigenous peasant early warning practices, and to the system of emergency preparedness adopted by the traditional farming system (a few works stand as exceptions here, notably Watts 1983).

The following probably exhausts the list of activities or combination of activities carried out by peasants in distress, and frequently examined in the general literature on the subject: diversifying the mix of assets held by a househould; reduced consumption and suspension of traditional observances; graduated disposal of assets; expansion of non-farm activities; seeking assistance or favour from relations; migration and sale of one's labour power (Faulkingham 1977, Hitchcock 1979, Dirks 1980, Greenough 1980, Watts 1983, Longhurst 1986).

Of the works listed here (and this is only a partial list, but one which includes a representative sample of different approaches; for more, see bibliography at the end) the last three are relatively more multi-faceted, but be that as it may, the interplay between the peasant community and traditional values on the one hand, and the process of selection of appropriate responses on the other is not seriously treated in any of them.

Among the group of activities noted above, selective asset disposal has been taken to be the most important, and so indeed it is. Among pastoralist communities, rapid diversification of assets, asset disposal, and on many occasions, raiding have been identified as critical survival techniques (Cossins 1972, Devitt 1977, Dahl and Hjort 1979, Toulmin 1985). For obvious reasons, we shall leave out of our account resistance and rebellion as a form of survival strategy.

Turning to Ethiopia, we find that the subject has been treated in several works, although—and this is an important point—not as a problem in itself nor in complete and systematic form. Most documents prepared by RRC (and a few by ENI) have a section or two on peasant migration, sale of livestock and rural grain price behaviour, each of which is used as an index for measuring the on-set or intensity of hunger in a given part of the country at a given time. In the same way, Cutler

(1984), and Seaman and Holt (1974) have dealt with aspects of asset disposal and migration as part of the effort to assess price behaviour during drought in northern Ethiopia in the former case, and the nutritional status of a population in eastern Ethiopia in the latter. A brief discussion of famine response is found in Gryseels and Jutzi (1986), but the main focus here is on the practice of selective livestock selling adopted by peasants in northern Shoa during the famine of 1985. Turton's work (1984, 1985) takes as its case material a tiny ethnic minority, the Mursi of Gamo Goffa, and identifies migration as a key element of the group's response to food shortages, and as an important factor in the relations between the Mursi and their neighbours. Crisis coping among another tiny minority, the Wayto, is dealt with in Teclehaimanot (1986).

On the other hand, Cutler and Stephenson (1984: 5–6) have attempted to draw what they call a famine response model which groups together most of the activities listed above and others related to market behaviour in a specific sequential order. The authors argue that each response should not be seen in isolation but as a series in a process. In contrast to the rest, this work provides a more rounded picture, but it contains some serious flaws which we shall discuss later in this study.

In the pages that follow we shall try to show that the survival strategy adopted by a particular peasant family (and we insist that the family continues to function as a unit even at its darkest hour) at a particular time is conditioned or influenced to one degree or another by the following: (a) The expected responses or strategies of other households in the community. By expected responses is meant responses that the individual peasant may reasonably anticipate others to make, or a set of actions agreed upon by the peasant and his neighbour(s). (b) The nature of the community, community relations, and accepted social values. A community is not made up merely of a group of individuals, there is a system of values and a body of moral obligations that bind these individuals together. A community which is more homogeneous is bound more closely together than another which is more heterogeneous.

Community and the ethic of cooperation

There is no community—at least not in Ethiopia—which is homogeneous or heterogeneous in the absolute sense. Absolute heterogeneity will in fact sever those delicate ties which define a community. Its opposite, absolute homogeneity will more often than not be of limited value, and, indeed, an indicator of social stagnation and atrophy. Homogeneity is considered here in two senses, namely the cultural/ethnic, and the social/class. The communities in which we did our field-work, and which were the main sources of our case material, namely those in Ambassel awraja were by and large homogeneous in both these two senses. We shall explain what we mean by community a little further on.

All the respondents in our questionnaire belonged either to the Moslem or to the Orthodox Christian faith, with those in the first group making up 84 per cent of the total. In Ambassel woreda, one of the three woredas of the awraja, Moslems in our sample made up 45 per cent of the total, while Christians 55 per cent, but otherwise Christians were insignificant numerically in the other two woredas, Worebabo and Tehulederae. Both Moslems and Christians spoke the same language, Amharic, exclusively, except a small number of peasants in the eastern corner of the awraja (statistically insignificant in our sample) who also spoke the Oromo language. Both religious groups were sedentary agriculturalists. We found very little evidence that ethnicity plays a significant role in community relations. Wollo has no history of religious fundamentalism, either of the Moslem or Christian variety, and there is a great deal of tolerance and social interaction (including marriages) between the two religious communities.

There is a high degree of social homogeneity not just in Wollo but throughout rural Ethiopia thanks to the radical land reform launched by the military government in 1975. There are two important aspects of this reform which are significant for our purposes here. First, in each rural area or region, land distribution was made to peasant households on the basis of a set of criteria (family size was the main one in many instances) uniformly employed during land distribution. The net result has been not just the abolition of all landed interests from the countryside but also the creation of a high degree of relative equality within rural society. Secondly, land reform has given rise to a continuous levelling down process tending in the long run towards the formation of a peasantry of diminutive holdings and of absolute equality. I have discussed both these issues in greater detail elsewhere (Dessalegn, 1984), and there is no need to dwell on the subject here.

The main point to note is that social differentiation within rural Ethiopia, which in the pre-revolution period was deep and many-dimensioned, has been replaced by relative social equality. This equality has in turn been accompanied by, or in some instances given rise to, great economic impoverishment. The impoverishment of the peasantry may also be explained in part by the fact that land reform together with subsequent rural policy adopted by the government has stopped peasant mobility, and closed many sources of non-agricultural employment and income. In Ambassel awraja and many areas of Wollo the most recent, and according to some extension agents the final, land distribution (called locally *Sir Neggel dillidil*, i.e. radical apportionment) was carried out in 1983–84.

It should be pointed out, however, that the seeds of social differentiation have not all been eliminated in rural communities. Some peasants have more farm animals and more and better tools than others. If family size was the main criterion employed in land distribution—and this was the case in Ambassel awraja and many parts of Wollo—a peasant with a large household will have acquired more land than another with a smaller one. And if the former peasant was also blessed with more and better quality supportive labour in the family, he will be in a better position to earn more income than his unfortunate neighbour. In both these cases, the advantages enjoyed are in large measure a result of demographic factors, and when the demographic contour changes, most of those who were in relatively privileged positions may not retain the same status.

But there are other causes of inequality, some accidental in origin, others a result of strength of character and skill in labour. By good luck or bad, some peasants received land in good locations while others ended up with plots in poorer areas. Great care was taken during land distribution to be as fair as possible to everyone, and each peasant was offered plots from each of the different categories of land in the community. Thus a peasant in Ambassel often works at least three different plots consisting of fertile, medium and poor land. But the physical location of a plot, i.e. whether or not it was on a mountain side, near to a road or footpath, far from or close to the homestead or a settlement, presented difficulties and had to be left to chance. Nevertheless, while good luck was of considerable advantage, it was the human quality which made the critical difference in the long run. The frugal, hard-working and skilled peasant was more likely to "prosper" than the peasant with less worthy qualities.

There is another factor which has to be taken into account when dealing with social differentiation, and relations between communities, and this has to do with environment and settlement. Certain ecological settings, by which we mean the natural environment and the economic and social consequences arising out of that environment, may be more salutary, and economically advantageous than others. The economic advantages obviously flow from the difference in the resource base of the settlements, but other factors also play a part, such as health (cf. Roundy), access to modern transportation, and location vis-à-vis important commercial or urban centres. Settlement and ecology of course determine product specialization and division of labour, and this factor will play a part in the relation between different communities.

It is worth emphasizing that the basis of social differentiation in rural Ethiopia since land reform is often not the size of land cultivated, but rather assets such as livestock and other movable property, traction power, and in a good number of cases the *labour potential* of the family, i.e. the number of able-bodied and working persons in the household. A peasant in a strong economic position is known in Wollo as *bal'e goulbet* or *goulbet yallew*, meaning literally one who has force or power. The term *goulbet* (force) refers to the extra labour power the peasant has in the family. In many parts of Wollo including Ambassel, labour and traction power alone—and in the latter case it is enough to own just a pair of oxen—were the most important factors giving rise to differentiation.

There is another, less tangible, factor which should be included in the discussion on the causes of differentiation, and this may be called enterprising spirit. The term refers to the resourcefulness, initiative and determination of a peasant in the pursuit of economic gains especially from farming-related or non-farm activities. Handicraft production is not an important source of supplementary income in Wollo, but trade in agricultural goods provides the enterprising peasant opportunities for more earnings and more "capital".

In Ambassel awraja a peasant maintained a strong position partly by renting out farm oxen, and partly by working the land of a poor peasant in return for a share in the harvest. These were the most important sources of higher earnings here, although there were a few peasants who were also active in trade.

It may be well to ask in contrast what the causes of economic misfortune are, and how these are manifested in concrete form. A poor peasant in most parts of Wollo is one who does not have sufficient labour or traction power. This may be due to several reasons: (a) The household may be small or there may be few able-bodied family members; young peasants who have only recently established their own homesteads, or older peasants whose children and dependents have for one reason or another left the family, fall in this category. (b) The family may have lost its livestock due to drought or cattle disease, or sold them off to make ends meet or to survive through hard times. (c) A poor peasant may also be in this position because he is advanced in years or is in poor health and depends on others to work his land. (d) Finally, women-headed households often fall in the ranks of the destitute as they too depend on others to cultivate their possessions.

We must point out, however, that the identification of who is poor and who is not is not as easy as the fore-going discussion may suggest. Take the following case which, incidentally, is not imaginary but true of one of our informants whom we interviewed in Wichalae, the capital of Ambassel woreda. The peasant has two oxen, but the family does not have enough labour, and the peasant himself (who is in his sixties) has been in poor health for many years. That farming season (i.e. beginning in July 1986) he rented out one of his oxen to his neighbour for a small fee to be paid in grain, and made arrangements to have his land cultivated by one of his in-laws. The relative was promised a small share of the harvest for his labours. His land, he said, measured approximately 0.8 hectares, not including the home plot. After his land had been ploughed he rented out the two farm animals to a second neighbour for a share of the latter's harvest. The community of this peasant had not been seriously affected by drought for several years. Now this peasant can by no means be described as a poor one by the standards of his community, although he described himself as such. On the other hand he was dependent on others for his livelihood, and it was this dependency which led him to consider himself as part of the unfortunate.

There is another aspect of social differentiation in post reform rural Ethiopia that needs to be discussed, and this has to do with what may be called the lasting power of status and of economic positions. In the pre-revolution period class positions were inherited by blood and marriage, and the landlord was certain that he would pass not just his wealth and property but also his class status on to his children. Land reform has changed all that. The insecurity of tenure brought on by the nationalization of land, and the general levelling down that has followed the reform has meant that a peasant's wealth now consists of "perishable commodities", notably labour power and livestock. This is the basis of the fragility of social positions prevalent in rural Ethiopia today where a "prosperous" peasant (using the term in its relative sense) may find himself thrown into the ranks of the destitute almost overnight due to some unforeseen event.

Consider the following case. Frost is a serious hazard in some parts of Wollo, and it is as much a cause of crop failures as is drought in these areas. A well-off peasant may get up one morning and find his entire harvest destroyed by frost. To survive through to the next harvest he may be forced to sell his livestock, perhaps including his draught animals. In the meantime his children or dependents decide to leave home either to set up their own households or to seek employment elsewhere. The economic position of our peasant will now be dramatically different from what it was before, and he may now have to turn to other peasants in his community who have escaped the disaster but who may have been less fortunate than himself earlier for assistance and support.

The degree of social differentiation varies considerably from one community to another. While in some the level of inequality may be measured in terms of a few heads of cattle, in others it may entail a gap in income of relatively wide proportions. Often enough, the better-off stratum of the peasantry consists of individuals whose main-line activity is not confined to agriculture but also includes rural trading. Trading in livestock is a lucrative business, and the peasants of southwestern Wollo (those in Borena and Wore-Illu awrajas especially) are well-known livestock traders. In the past, energetic Wollo peasants used to travel long distances—to Kaffa and Illubabor provinces, for instance—to work as seasonal labourers and to deal in cattle and grain; this is no longer possible now. Grain trading is also a source of supplementary income, though on a smaller scale, and large grain merchants in Wollo often tend to be town dwellers rather than peasants.

All things considered, however, it would be a mistake to argue that in Wollo as in many other parts of rural Ethiopia the social and material conditions exist for the evolution or consolidation of fullfledged classes in the Marxist sense of the word.

So much nonsense has been written about rural class formation in Africa that the subject needs to be carefully re-examined, and some of the sound and fury regarding the alleged capitalist offense in the African countryside be given the burial it deserves. This, however, is not the place for such endeavours, and we will have to content ourselves with only a few words on the subject.

The most serious fallacy in the literature, in our opinion, is the widely held view that rural Africa is split into three, often warring indigenous *classes*, labelled poor, middle and rich peasantry. The categories themselves are incomprehensible: consider, for example, poor proletariat, middle proletariat and rich proletariat as distinct class categories; see e.g. Bernestein 1979. These same "classes" are often said to be subordinated to urban-based capitalist forces, or to multinational agribusiness and international capital.

Some proponents of this view have also argued that rural class formation and the consequent domination of capitalist forces and relations in the countryside are to blame for food shortages in many parts of Africa (Cliffe and Moorsom 1979, Shoeph 1985, Shenton and Watts 1979). One writer attributed the cause of the 1973/74 Ethiopian famine to the "shattering impact" of capitalist development (Cliffe 1974) —surely an undeserved insult to capitalism; one wonders what other sophistry he will come up with to explain the famine that occurred ten years after "capitalism" had itself been shattered.

Here in this country fulminations against an alleged *Kulak* class, which was accused of attempting to sabotage the resolution in the countryside, by state authorities and individual writers (Galperin 1981) went on for several years until the Kulak idea was dropped as mysteriously as it had originally been taken up. In our view, the identification of rural classes must be based on the following considerations:

1. Whether or not a cultivator operates his farms predominantly (not necessarily exclusively) for the market and for profit, and whether what he produces are commodities and not self-consumed use-values. The peasant who goes to the local market to sell one crop in order to buy another is not engaged in the market economy. This kind of exchange has a history stretching over many centuries.

2. Whether or not the person employs *wage labour*, which must not be confused with family, associated or communal labour (*see* the debate on the subject in Long [ed.] 1984).

3. Whether or not he employs techniques of production relatively superior to those used by the surrounding population engaged in agriculture.

The foregoing discussion notwithstanding, the social stratification evident in rural Ethiopia cannot be ignored, for, among other things, as we will see later, it has played an important role during the food crisis.

To admit the existence of rural stratification but deny that of social classes may appear contradictory, but we believe it is closer to the existing reality to consider the peasantry in Wollo as a single class, but differentiated into several strata. In this work, the term peasantry includes all rural producers except those who live predominantly by pastoral production. The agricultural "proletariat" in Wollo (as well as in the rest of the country) is such a puny force that it is best to exclude it from consideration.

We shall distinguish three levels of stratification within the Wollo peasantry (the terms are descriptive): poor (or lower) peasants, self-supporting (or middle) peasants, and upper peasants. The last group should not be confused with "rich peasants" often discussed in the literature on rural class formation, for upper peasants are not capitalist-oriented but merely earn more than the others. Some might point to a fourth group, i.e. marginalized peasants who often have no land of their own, earning their livelihood by providing labour, and who in most cases are dependents in other peasant households. This group is difficult to identify statistically (because it is part of other peasant households), but we believe it is small in number as many in the group often migrate to the urban areas and become part of the urban unemployed. For analytical purposes peasants on the margins are better included in the lower peasantry.

For reasons to be discussed a little further on we were unable to collect proper quantifiable data related to income, and economic status, and hence we are not in a position to measure accurately the numerical strength of each group. However, on the basis of evidence available to us, we estimate that 50 per cent of the peasantry in Ambassel awraja fall in the ranks of the lower stratum, another 40 per cent in the middle, and 10 per cent in the upper stratum. These figures are estimates based on indirect evidence and should not be taken as definitive. Since land reform, and as a result of the many crop failures that the area has suffered in the last ten years, the lower peasantry has grown in size considerably and will continue to do so in the years to come.

The relationship between upper and lower peasantry, like any relationship based on relative inequality, benefits the former more than the latter. Nevertheless, the relationship should not be taken to be "exploitive", that is, one involving the appropriation of surplus product or of labour in the Marxist sense. In many instances, the relationship entails mutual benefits to both sides, and in some it may be the better-off peasant who expends the necessary labour rather than the weaker one. Let us illustrate this with an example.

As suggested above, it is not uncommon in Ambassel awraja as well as the rest of the province for a poor peasant (one who does not have draught animal power, or who may be unable to do strenuous physical labour, etc.) to rent out his land to someone in a better position than himself. The latter agrees to farm the land of the poor peasant in exchange for a portion of the harvest. How much the share of each peasant will be in this arrangement will depend on who has contributed the seeds, and how much ancillary labour the landowner himself has provided during the period of weeding, harvesting, threshing, etc. In this particular arrangement both sides stand to benefit, and it may even be said that it is the poor peasant who "appropriates" the labour of the strong one.

Let us now turn to the concept of community. In most parts of Wollo peasants do not live in compact villages but in dispersed settlements, although the distance between homesteads may not be great in all cases. (This may soon change as the government stubbornly pushes ahead with its villagization programme.) However, most households belong to individual communities each of which reveals subtle marks of distinctness. For the purposes of this study we shall identify three levels of community: the *core*, the *kebbelae*, and the *local* community.

A community is basically a set of diverse relationships sanctioned by commonly received values. A fine illustration of what is meant here is the relation between the highland and pastoral lowland communities of Wollo, especially during periods of drought and high cattle mortality. The practice of raiding cattle, until recently quite common in eastern Wollo, was socially sanctioned across community border lines, i.e. between farmer and herder, and each community acknowledged it as legitimate against the other (Cossins 1972: 43ff). Raiding the property of a member of one's own community, however, was strongly disapproved of. The distinction we have drawn among the three levels is based not only on the spatial arrangement of the communities but also on the fact that as one moves from the inner to the outer, the content of relationship subtly changes.

The core community may be likened to a parish, and consists of about one hundred or more households located within reasonable distance of each other. Such a community is known as *got* in Wollo, and until the advent of peasant associations (PAs) the *got* was the most important social frame of reference of every peasant here. Even now many peasants in the region—highlanders more than lowlanders—identify with the *got* rather than with the larger *kebbelae* or peasant association. A *got* may contain one or more *mender*, which is a settlement composed of a cluster of homesteads resembling a hamlet. Rare is the *got* where the population is made up of divergent elements; in most cases the people are of the same cultural background.

Several *got* form a *kebbelae* which at present falls within the boundary of a peasant association (PA). Although the *kebbelae* and the PA, to which all households are expected to belong, are taken to be one and the same, the original and deeper concept of *kebbelae* is much broader than the PA. In this work, however, we shall take the two to be more or less identical.

The numerical strength of PAs is much greater in Wollo than in many parts of the country. The average PA in Ambassel awraja, for instance, has a membership of 760 households with a population of about 3,200; but a good number of PAs have a membership of over 1,000 households and a population of over 5,000. There is a considerable degree of cultural uniformity within each *kebbelae*, although it is not uncommon to find communities consisting of groups with different backgrounds. There was, relatively speaking, greater heterogeneity among PAs in Ambassel woreda than in the other two woredas of our awraja.

What we have called *local community* is larger in size, and more diverse socially and ecologically than the previous two communities which are included in it. Locality as a loose but definite social and economic relationship, and a specific mix of micro-environmental settings, may be difficult to define spatially but it is a valid concept and a frame of reference often employed by peasants themselves. A local community embraces several peasant associations, perhaps as many as half a dozen or more, and may on occasion cut across administrative boundaries.

Depending on its physical size, population density, topography and agricultural system, a woreda may contain four or more local communities of this sort, some distinct enough to be identified by their own names, others less outwardly distinct but nonetheless functioning as communities. In this first category may be settlements situated in distinct physical settings, such as plains, plateaux and the like. Thus, for example, the settlements or peasant associations in the Gerado and Sirinka plains, the one west of Dessie, and the other in southern Yejju form easily identifiable communities.

Very often a local community will form itself around the woreda capital, or at key point on a well-travelled road. Wichalae is not just the capital of Ambassel woreda but around it is a local rural community which frequently is referred to by the same name; this particular community actually extends itself across the border into Yejju awraja. Travelling south to Haiq, the capital of our awraja, one will pass at least two local communities, the first around Tis'abalima (now increasingly dominated by the cooperative established there), and the second around Golbo to the west.

There is more diversity at this level than at the other two, partly because the community is larger and embraces more people, and partly because the relatively greater micro-ecological variation gives rise to relatively greater specialization and division of labour. Here, exchange relations play a greater role as relationships take on "impersonal" aspects, and this is symbolized by the fact that most local communities possess at least one important market centre which attracts visitors from many other communities each week. Quite often such market centres are renown for trade in one or two agricultural products, such as cattle, crops or one specific crop, dairy products, etc. Wichalae, for instance, is a good market for grain, especially oil crops, and Golbo is well-known as a cattle market.

At each level of community one observes two important forms of relationship, the one personal, the other economic. In the first category are relationships based on kinship, marriage, friendship and, at times, religious ties, and in the second those having to do with economic arrangements and exchange. It is often the case that the personal and the economic operate jointly and in a manner that reinforces both relationships. A peasant decides to sell a young bull to another peasant because both are related by marriage; he prefers this rather than the alternative of selling the animal to a stranger because he can count on borrowing or renting the same animal for farming a few years hence should his fortunes change for the worse.

Men and women often attempt to spread their personal relationships as widely as possible, often through marriage and other personal ties, as a form of insurance against hard times; this involves forming relationships not just in one's own community but in others. Not infrequently, localized, small-scale disasters affect one community and not another next door, or at times only a part of a community with the other remaining unscathed, and a peasant in dire straights but with friends or relatives untouched by the disaster will recover less painfully than another without these kinds of ties.

During the course of a food crisis—and taking into account the pre-and-post phases of it, the time span of a crisis in Wollo ranges from nine to fifteen months or more—all forms of relationships at all levels of community come into play in a heightened manner and at an accelerated tempo. The rural world is turned into a scene of feverish activity as men and women attempt to take protective measures and make short- or long-term arrangements to insure the survival of their families. Neighbours and friends decide to pool their resources the better to withstand the hardship; agreements are reached between relatives or friends to dispose of assets in turns, and to support each other in the meantime; measures are taken to remove livestock to areas in the community or to other areas which are less exposed to the crisis and to leave them in the care of acquaintances or fellow peasants with or without compensation involved; arrangements are made to sell livestock to peasants in one's own community or a neighbouring one with the understanding that at the end, the sellers will rent the animals for farming purposes; markets both in the neighbouring communities and in distant ones, especially those reported to be relatively free from social or ecological stress are frequently monitored, and the information disseminated widely; distress signals are sent out to relatives living in urban areas or in other awrajas or provinces.

The list may be extended, but it is sufficient for this conclusion to be made: the flooding of the market with livestock, and the dramatic drop in the price of cattle, and the mass migration of peasants out of the rural areas that one observes at a particular point in a famine are merely the closing scenes of a drama whose most important and most decisive acts have already been played out behind closed curtains, as it were.

Guiding all of these activities and informing the relations of families at all levels, is the common set of social and moral values shared in the community of which the most important for our purposes is that which we call the *ethic of cooperation.* Simply put, cooperation involves acts among individuals in which the pursuit of mutual support and reciprocal benefit—not always and necessarily equally shared—is the major element in the motivation behind the acts.

It is often suggested in the literature that crisis, especially food crisis, brings out the worst in men and women. But this is a one-sided, and at times distorted view often arising from the practice of looking at the problem from the standpoint of what disaster does to its victims. Thus people speak of husbands driving away their wives and children (or of deserting them), of members of the same family locked in battle for food, of cruel and abnormal behaviour, of mothers eating their children and other acts of cannibalism (see Greenough 1980, Pankhurst 1985). There is, however, sufficient evidence for an opposing argument, namely that in a good number of cases famine brings out the best in individuals caught as its victims. The spirit of cooperation, of sharing assets, resources and services, and of reciprocal support shown in rural communities during food shortages plays a critical role in peasant survival strategies.

Here are the main types of cooperation that underlie peasant response to hunger and hunger-induced crisis. It is common practice in Wollo for two peasants with one draught animal each—and in crisis conditions peasants have been known to use donkeys and cows as draught animals—to team up and work their land in turns. This is the simplest form of cooperation, one which is based on equality of resources and benefits. This practice will have two implications: (a) It will have a bearing on the decisions of peasants related to the selective sale of livestock and other assets during the course of famine. (b) Post-famine recovery would be more difficult than it is if this mode of cooperation was unknown. It is also common practice in the region for a peasant with enough or surplus traction power to lease his work animal or animals to someone in need, to be rewarded for the service with a portion of the harvest.

There is another side to this form of cooperation, and that is that a peasant with sufficient draught animal power will, for an agreed upon share of the harvest, work the land of someone unable to farm his holding. There are other variations to this relationship between those with traction power on the one hand and these with only land, but all are forms of cooperation even though remuneration is provided for the service given. To appreciate how important and mutually beneficial this form of cooperation is considered, what would happen if the practice was unknown? Here again, the fact that this sevice is available will affect peasants in their decisions regarding asset disposal.

A third form of cooperation is communal or often group cooperation and is known as *jiggae* in Wollo. Here a needy peasant is helped to work his land, build his homestead, and generally acquire labour and other inputs (including seeds) by a group of friends, relatives and neighbours to enable him to get on his feet. The peasant is expected to reciprocate when another peasant needs this kind of assistance. This traditional form of group cooperation is employed frequently in post-famine recovery, but it is extended to involve the community as a whole and may be viewed as communal cooperation. Here peasants in a community, usually the *got*, mobilize all their resources and work the land of all peasants in need.

There are other forms of cooperation peculiar to peasants in Wollo province. An interesting practice which a number of our peasant informants in Ambassel and neighbouring awrajas told us about is what may be called *sharing of homesteads*. As the famine intensifies a family may decide to sell its home, that is the building material, and move in with a friend or relative. The home is sold piecemeal, first one part, then another, until the whole thing is stripped clean, and part or most of the proceeds are shared with the host family. The guest family will stay with its hosts (unless the famine drives them to sell their home, too) until it is strong enough to rebuild its homestead.

Another practice reported to us by peasants here may be called *sharing a 'goudguad'*. This practice, not uncommon in the years before the revolution, is now becoming rare, as a result of the general impoverishment of the peasantry. A *goudguad* is a pit dug in a convenient corner of the homestead for the storage of grain. How big a *goudguad* is will depend on how big a harvest a peasant usually expects, and how much surplus he retains.

In the past, a prosperous peasant was able to own more than one *goudguad*. Nowadays it is only the top echelons of the upper peasantry that are capable of retaining a "surplus" in their *goudguad* during a prolonged drought or successive crop failures, and it is said that a peasant with a good-sized *goudguad* can survive famine for up to two years. On occasions a peasant with a "surplus" will strike up an agreement with a number of peasants who have exhausted their stock of food to share his *goudguad* and to replenish it as soon as their fortunes improve. The owner of the *goudguad* will be compensated later because each participating peasant will pay back in kind a little more than what he took out in the arrangement.

Finally, when a peasant family decides to leave the community and migrate in search of employment or assistance, its land (and other possessions if it has any) are left in the care of one or more peasants who will be entitled to use it in their absence. If the land of the absent family is successfully farmed while they are away the family will have a small stock of food waiting for them when they return.

Some may see in these reciprocal acts evidence of the importance of moral relations in peasant society or the workings of what James Scott has called a moral economy. Certainly the ethic of cooperation is expressed in moral language, and reciprocity is believed by peasants to have the sanction of religious teaching. But the moral is a cover and justification for the material and the economic. At the heart of most forms of peasant relations is the need to assure the livelihood of the household and the well-being of its members. No peasant has all the resources to independently withstand a major crisis, and hence reciprocal support is sought by everyone. Cooperation entails benefits to all parties involved, although, as was noted earlier, some may benefit more than others.

The spirit of cooperation begins in the household, and radiates, as it were, outwards. The family is an important unit in peasant relations, and, as we shall try to show later, it does not disintegrate quickly or easily in the face of hardship. Furthermore, the household is not turned into a battlefield in which members fight each other for the allocation of food, nor are the weaker members victimised by the stronger when starvation begins as has been argued in some of the literature (Sen 1983, Greenough 1980, McCann 1986 ms.). Rather, in a good number of cases attempts are made on the part of each member to preserve the integrity of the household.

It will be clear from the foregoing discussion that peasant survival strategy is best understood as a collective endeavour. What may appear to the casual observer as a disordered and isolated series of movements by individual peasants during the onset of famine is actually an ordered and group-centered effort to minimize the impact of the crisis and to stay alive. To be sure, if the famine continues beyond a certain point in time, and a certain level of intensity, peasant survival strategy will collapse, and with it all the individual's group-based plans and arrangements. This point, which is part of what we shall call the phase of death and dispersal, coincides with the complete exhaustion of all the resources of the community, including its alternative sources of food.

Before we close this chapter, a few words about the author's primary intentions in embarking on this research. The writer is convinced, firstly, that famine can only be defeated if all the resources of the rural community are mobilized and rationally and judiciously employed for the benefit of that community well before the harbingers of hunger and disease have begun to show themselves; and secondly, that emergency relief, which in this country means massive shipments of food aid from the west, may help temporarily to abate famine but will not remove the root causes of the problem, and will in fact lead towards greater dependency on external powers.

On the other hand, the rural population, especially the peasantry in areas with a long history of food shortages, has over the decades evolved relatively effective techniques of coping with hunger and privation, and if these techniques are improved, strengthened and extended, they can form the basis of a viable and indigenous system of famine survival. This is the most cost-effective and the most easily diffused strategy available to this country. The task of freeing ourselves from the scourge of famine as well as from external dependency must begin by first taking stock of the country's capabilities, potentials and deficiencies. The first order of business in an endeavour of this sort will be to examine closely the "science" of survival of the peasantry in all of the country's distinctive agro-ecologies and systems of rural production, and to evaluate the merits and demerits of each. It is hoped that this small work will serve as a modest contribution to that effort.

2. Organization of the Study

The study is broken down into four sections. Section II attempts to situate Wollo in its ecological setting, and to provide an extended discussion of peasant agricultural practices there. A few pages are devoted to the general economy of the province, but since very little work has been done in this area, the discussion is incomplete, and unsatisfactory to some. In Section III the main subject of our work is discussed, i.e. the indigenous survival techniques of the peasantry and post-famine recovery which reveals the scars and wounds of the struggle for survival. The final section has two main aims, first, to review briefly some of the arguments having to do with famine in the general literature and to examine their relevance to Ethiopia, and, secondly, to assess the lessons to be drawn from the experiences of the Wollo peasantry, and to evaluate their implications for policy.

The material collected for the study in Wollo is primarily of two kinds: first, information and documents about the province as a whole, and, secondly, field data and official records collected more systematically about Ambassel awraja, one of the twelve awrajas of Wollo; this to serve as a specific case material for the study. The organization of the analysis is similarly bimodal, as it were: the discussion will, in many parts, move from the province to the awraja and back to the province alternately, with primary evidence and supportive data provided for the most part from the findings of our field enumeration and record collection in the awraja.

Why did we select Ambassel and not any of the other awrajas of Wollo for the field work? Of the several reasons that affected our choice of a field site, the following two are the most important: (a) In many ways, social, economic and ecological, the awraja is a reflection in microcosm of the province as a whole. A rugged terrain in the highlands to the west, flat plains in the middle, and desolation in the lowlands to the east: such in brief is the physical profile of both province and awraja.

(b) As in the province as a whole, so too in the awraja, some parts were untouched or lightly affected by the 1984/85 famine, while others were devastated by the disaster. It was important for our study to examine the relation between the people of these two regions during and after the crisis. In addition, Ambassel was the most easily accessible of the major famine-prone awrajas of Wollo, and this was an important factor as we were operating with a modest research budget, a small staff (consisting of this writer, his field-assistant and a driver), and a tight schedule.

The Sources for the Study

The primary sources of information for the study were four in kind.

(a) *Structured interviews of peasant household heads*
A carefully designed questionnaire consisting of thirty interconnected questions was employed to collect quantitative information. The questionnaire was administered on a selected and representative sample of peasants from all three woredas of Ambassel awraja, i.e. *Ambassel, Worebabo and Tehulederae* woredas. (Note that Ambassel is also the name of one of the three woredas of the awraja. Hereafter "Ambassel" used alone refers to the awraja.) The sample size was originally set at 300 (100 for each woreda), but for reasons discussed further on the number had to be reduced to 255. For the sake of uniformity and consistency the interviews, conducted for the most part by my field-assistant, were completed in one operation in the weeks between 18 October and 5 November 1986.

Researchers who have had field survey experience know how difficult it is to get hold of the right kind and number of peasants for interviews, especially where, like in Wollo, the great majority of homesteads are inaccessible by road, and settlements are spread out over a wide area. All care was taken to make the sample of respondents from each peasant association (PA) or woreda as random as possible, but perfect randomness was not possible to achieve. To begin with, the questions in our questionnaire were meant to be put to peasants who had to varying degrees personally experienced the 1984/85 famine. Those peasants whose communities had escaped the disaster, and who had not therefore suffered privation as a result, were not selected for the interview. Secondly, peasants who lived too far away from the woreda capital or extension centre, where the interviews were conducted, too ill to travel, or unwilling to be involved in the interview altogether were likewise excluded.

In each instance the areas and PAs affected by the famine were identified for us by the chief extension agent of the woreda. This information was immediately checked with officials of the woreda peasant association (WPA) from whom a roster of members of each organization was obtained (the offices of the woreda extension agency or MOA and WPA are found close together in almost all parts of Wollo). In each PA thus identified, a select number of household heads was picked out randomly for the interview. Peasants who did not show up for interviews or who declined the request were immediately replaced by others from the PA roster, which was up to date having been prepared shortly after the 1984/85 resettlement programme had been concluded.

At the time of our field-work the peasantry in many parts of Wollo was still receiving emergency relief food which was distributed by RRC, the Ethiopian Red Cross, and other agencies. In Ambassel, the distribution was done on certain days in the month on which peasants from a specified number of PAs were scheduled to travel in turns to the woreda capital or extension centre, where the food was stored, to pick up their rations. Fortunately for us our visit to Ambassel and Worebabo woredas coincided with the distribution schedule of the PAs selected in our sample, and all we had to do was wait in the woreda capital for "our" peasants (in Tehulederae, however, we had to travel to many of the PAs in our sample). This saved us a great deal of time, effort and the frustration that is often inevitable in field survey work. If we had carried out the interviews at each peasant association it would have meant many days of travel on foot and by mule, and the work would have taken us not the two and a half weeks that it did but no less than two and a half months. The emergency food distribution was scheduled to be terminated in December.

Now this stroke of luck unfortunately led to an unforeseen outcome which we realized only after we had completed a good part of our work. It was not necessary for a household head, in whose name the food is provided to the family, to personally appear on distribution day to collect his/her family's ration. He or she could send someone in the family, or delegate a neighbour to pick up the food instead. On occasions, one person with one or two pack animals (sometimes borrowed) would be entrusted with the task of collecting the food of his neighbours (for a small fee). Each adult was allowed 15 kg of food, and a family of four persons (two adults and two minors) received up to 50 kg of food, and at least one can of cooking oil each month. Now this was too much to personally carry back to the house particularly for the elderly. Women household heads were often elderly widows, and the monthly travel to and from the distribution centres, which might take up to six hours or more, was a burden which many often avoided either by sending a younger member of the family in their place, or by entrusting the job to a friend or relative.

The result in short was that the number of women that appeared in our questionnaire was proportionally far less than the number of women registered with PAs, especially in Ambassel and Worebabo woredas where much of the interviewing was done in the woreda capitals and extension centres. In the end this would have distorted the gender-oriented data we hoped to gather, and for this reason we had to make some changes in our plans, and to collect instead "soft" information (as opposed to hard data) on aspects of gender through oral discussion with women peasants in and outside our sample.

There are two other points which should be raised regarding the content and distribution of the field interviews and the questionnaire. Ambassel woreda was the least affected of the three woredas of the awraja, with only about 30 per cent of the PAs in it having suffered hardship, whereas the comparable size of victims in the other two woredas ranged between 40 per cent (Tehulederae) and 60 per cent (Worebabo). We therefore had to make adjustments in our sample to take account of this fact and to balance the impact of the first woreda vis-à-vis the others. We should also note in passing that a few of the completed questionnaire forms were found at the time of editing to contain errors and had to be discarded.

In selecting the questions to be included in the questionnaire we had to take into account the relief operation that was still going on, and the great material loss the peasantry had so recently suffered. Briefly, the relief effort was not restricted to the distribution of food rations but also included the distribution of seeds, oxen and agricultural implements to peasants who were considered particularly destitute. The NGOs rather than RRC were most active in this form of rehabilitation work. The Ethiopian Red Cross Society (hereafter referred to as the Red Cross), for example, had identified 260,000 households in Ambassel and Qallu awrajas as beneficiaries of its rehabilitation programme and had, as of October 1986, distributed 3,500 tons of seeds, some 2,500 farm oxen, and over 118,000 hand tools to needy peasants in the area. The Ethiopian

Orthodox Church, active for the first time ever in rehabilitation and rural development work and officially registered as an NGO, had "adopted" Yejju awraja and was distributing oxen there at the time of our visit.

Needless to say, a great number of peasants had lost livestock and were short of draught animal power, and recovery in this area was a long way off. Furthermore, peasants were still burdened with taxes, AMC-imposed grain quotas, and a host of other obligations to the state. It is sad but true that since land reform the peasant is weighed down with more and greater obligations—this time all to the state—than he was before. Under these circumstances, we felt it would be injudicious to ask questions related to peasants' economic status before or after the famine. Questions having to do with possessions, income, and "capital", especially livestock had therefore to be excluded because we believed our respondents would either provide wrong answers, or would take offense which would have badly affected their response to the other questions as well.

This is a pilot study, and it is not unusual for such studies to kick off with limited objectives. Given the constraints noted earlier, the scope and coverage of the research had to be kept within the resources available to the project. Despite these limitations, however, we believe our findings provide valid support for the analysis presented and the conclusions reached. Table 1 shows a breakdown of the sampled peasants by age and location.

(b) *In-depth discussions with peasants and peasant leaders*

Just as important as the data we gathered through the questionnaire was the information we collected through extended and unstructured discussions with individual peasants and peasant leaders. These num-

Table 1. *Respondents by age and woreda*

Woreda	Age Distribution					Total
	Up to 29	30–39	40–49	50–59	60 +	
Ambassel	–	6	18	24	17	65
Worebabo	4	13	23	23	37	100
Tehulederae	4	15	20	20	31	90
Total	8	34	61	67	85	255
%		16.5	24.0	26	33	100

bered some sixty-five of which two-thirds were in Ambassel and the rest in the surrounding awrajas of Qallu, Dessie Zuria and Yejju. It was not just information, i.e. facts and figures, names and dates, that we obtained from our dialogue but also insights, thoughts, attitudes, beliefs, and practices concerning many aspects of peasant life. If one may be allowed a sideline remark here, there is a tendency in many quarters in this country to make a fetish of the quantitative approach, and to see in the pat formula and neat mathematical model the answers to all questions and problems. But however perfectly formulae and models may be designed they will only provide a small glimpse of the rural world and a crude picture of peasant conditions. The face-to-face discussion and the invaluable observations and commentaries it brings out when well-executed is indispensible, and one hopes more and more researchers will give this approach the importance it deserves.

It is one thing to hold oral discussions with peasants and another to be fully rewarded by it. Not all peasants are willing to share their thoughts and experiences with strangers particularly during times of uncertainty. Peasants in Wollo felt ill at ease and somewhat insecure because of current government policies having to do with resettlement, cooperativization and villagization. Not many peasants are adept at expressing complex thoughts or deep experiences in clear and ordered form; verbal communication in rural communities often restricts itself to the bare essentials and the externals. The researcher himself may be unable to ask questions relevant and meaningful to his informant, and without the right queries a discussion will soon deteriorate into irrelevancies and trivia.

The quality of the information provided by an informant may be conditioned not just by the sensitive nature of the subject but also by several other factors. Whether the interview is done in private or in groups; whether or not local officials are involved, or present in the area where the interview is taking place; whether or not the informant expects a favour from the researchers in return. Peasants do not as a general rule expect to be remunerated for holding interviews, but on occasions they asked us favours, the most frequent of which were rides in our vehicles. There is something fascinating about vehicles for peasants, and some will climb in cars, on pick-ups or trucks and travel long distances out of their way just for the joy of the ride. At other times peasants asked us to take messages or pleas to higher authorities on their behalf.

In contrast to our experiences with peasants in other parts of Ethiopia, we found the peasants of Wollo not only open and friendly but quite willing to engage in discussions with outsiders. Other peasants in other parts of the country are often reserved, uncommunicative, suspicious and on occasions downright hostile. The peasants in the parts of Wollo we covered, on the contrary, were easy-going, humorous and almost voluble. This contrasts sharply with the experiences of Cossins (1974: Zone 2, 9ff) who found the peasantry of northeast Wollo secretive, sullen and hostile.

There were no constraints in our movements, and as we travelled from one locality to another we observed peasants at work: weeding, harvesting their crops (mainly barley), guarding their fields against rodents, turning the soil and building protection against soil and water erosion, constructing their homes or making repairs to them, etc., etc. The time of our field work, October and early November, was not a time of peak labour, or labour demand, which occurs between November and December, the main harvest season for most crops, and June–August which is the ploughing and planting period. Some of our conversations with peasants took place while they were taking a break from work.

(c) *Interviews with local officials, and records from government agencies and NGOs in the area*

Our main source of information in this instance were agents of the Ministry of Agriculture, especially those at the awraja and woreda levels. But we have also benefitted by the informal discussion we had with agricultural experts at the provincial office in Dessie, with officials of RRC in Ambassel awraja, and of the Red Cross in Dessie and Bati.

Another important source of information in this category were unpublished records that were made available to us by various government and non-government agencies both in Dessie and Ambassel awraja. Here again, the records we obtained from the Ministry of Agriculture offices proved to be invaluable. A list of the agencies from which we obtained unpublished records is provided in the first section of the bibliography at the end of this work.

(d) *Published Primary Sources*

Section Two of the bibliography lists all the primary published sources that we had access to and that were relevant to this study.

The quality of documents prepared by various government agencies in Wollo (as well as in Addis) is very uneven. There are a number of official reports about the famine in Wollo and elsewhere prepared by RRC and other government agencies at the national and provincial levels, but official reports in this country in general are notorious for their lack of detail, inconsistency, and obfuscation of significant problems. This is mainly because most reports are prepared hurriedly, and the authors often are bureaucrats rather than professional researchers or academics.

Those who have done research work in Ethiopia know how frustrating it is to work with official unpublished records and government publications. The data provided in these sources are often incomplete, the methods or categories used to compile the information change from one period to another, and there are—frequently—inaccuracies, misleading statements, and on occasion downright falsehoods. On top of this, documents issued by two government departments about the same subject almost always carry conflicting data even when one of the departments is the sole source of the information used. Here is a good example. Some records of the Ministry of Agriculture, Wollo, give a figure of 9.3 million hectares as the total land area of the province, of which about 10 per cent is said to be cropland area; other records of the same agency give a slightly different figure. The Planning Office in Dessie, basing itself (it says) on records from the same Ministry gives a figure of 6.5 million as the total land area of the province, of which 65 per cent is pastureland and about 30 per cent cropland (NEEPR, 1985: Part I, 2ff). Pastureland here is taken to include the Afar depression, which is a barren stretch of land containing water and grazing potential only in a few places. This area is estimated by some Ministry of Agriculture records as making up one third of Wollo. Using the Planning Office's data one arrives at a figure of 2 million hectares as the crop land area of Wollo, which is about four times greater than the figure provided in CSO's annual agricultural surveys (CSO 1984).

Needless to say, we have also made use of secondary material, not, however, so much as direct sources of information but as inputs in the analysis and debate presented in the work. The current literature on famine is dominated by works on Ethiopia. In a recent bibliography on famine in Africa, Ethiopia was featured prominently, taking up almost half the country listing (Seeley 1986 B: 15506). But much of this literature is unimpressive, a good deal of the works having been written hurriedly at the time of the famine.

Technical problems and usage

To avoid cluttering the pages we have kept references in the text to a minimum compatible with standard academic practice. Readers who wish to pursue a line of argument or of thought further and require the relevant sources are welcome to examine the reference section which, though by no means complete, contains a sufficient sample of works on famine and famine survival.

The spelling of Ethiopian place names in English continues to be bedevilled with inconsistency and surrounded with confusion. The more so now since some fools decided of late to rush in where formerly many angels had feared to tread. Unconcerned about the great difficulties that this would create, and virtually on their own initiative and without sufficient expert advice, several government agencies—the Ethiopian Mapping Agency, CSO, to name but a few—have decided to adopt their own rules for rendering place names in English. This has led towards greater anarchy, and has badly damaged the chances for an early solution to the problem.

The name of the province which is the main focus of this study is now spelled in at least three different ways: Wallo, Welo, Wollo. Its capital is rendered in three different ways: Dase, Dese, Dessie. In each case, the older and the more widely used (at least in the earlier literature) is the last spelling which we have decided to adopt in this work. We felt that until the confusion is settled once and for all—as things stand now this is unlikely to happen for several decades—it will be less trying and easier for all concerned to employ the older and widely diffused spelling of place names wherever possible.

But the problem does not end here: what about the numerous local places which are often referred to in this work, some of which do not even appear in maps prepared by the Mapping Agency? How are these to be rendered in English? I have used my own judgement here, and have transcribed the local names into English as best I can. The two maps (Maps 1 and 2, p. 43 and p. 67) have been prepared using the older spelling for the larger places, and my own approximation for the smaller ones.

There is another problem to clear up before we can turn to the subject at hand and this has to do with the usage of certain terms in the text. The administrative division of the country consists of three basic units called in Amharic *woreda*, *awraja*, and *kifle hagger*. There are no official guidelines as to how these terms should be rendered in English, but while the first two have often been retained in their Amharic original, the last has been translated as "administrative region" in recent years, although in the past the term often used for it was "province". To complicate matters some zealous officials have now begun to render "awraja" as "province", and this has found its way into reports, documents and publications.

We believe the most appropriate term for *kifle hagger* is "province" as "administrative region" is cumbersome (try for instance the adjective, like "provincial transport"), confusing because there are now planning regions, military regions and resources regions, and technically incorrect. In this work we shall retain the older usage: we shall speak of Wollo province, Ambassel awraja and Worebabo woreda.

A final problem of usage has to do with the names of some government agencies. The current craze within the Ethiopian bureaucracy may be called "zonalization". The National Committee for Central Planning has created its own planning regions, and established several zonal offices in the country, including one in Wollo. The Ministry of Agriculture has a different zonation system, and it too is now opening up zonal offices of its own. Its office in Dessie, the capital of Wollo, is now called the northeast zonal office of the Ministry of Agricultural Development. Apart from the terminological confusion this will give rise to, the main damage will lie in the fact that yet another layer of bureaucracy will be tacked on to a governmental structure already of elephantine proportions by many standards. For ease of reference and simplicity we shall continue to use the term Ministry of Agriculture (MOA for short), followed by the name of the province, awraja or woreda if it refers to a specific branch office.

II FAMINE: HIDING BEHIND THE MOUNTAINS

3. Wollo and Ambassel: The Setting

The most striking feature of Wollo is its rugged topography. The province is dissected by high mountains and mountain chains, deep ravines, broad gorges and numerous rivers and streams. The great escarpment, stretching north to south, runs through the middle of the province dividing it into two halves. West of this is mountain country, with peaks rising well over 3,500 metres, and high plateaux reaching into the alpine zone; east of the escarpment the land falls steadily, with here and there high plateaux, until it merges with the Afar lowlands. The western borders of the province are bounded by the Abbay-Beshilo-Tekkezae river gorges, and in the south the terrain becomes an extension of the northern Shoan highlands. On the whole, the area is drained in two directions: streams and rivers from the rugged highlands unload their contents into the Abbay-Beshilo-Tekkezae river systems in the west, while in the east the Awash river receives the waters of numerous streams (see Map 1).

In a normal year, many parts of Wollo experience two rainfall seasons, and a good number of peasants follow a bimodal farming system. The two seasons are known as *belg*, which falls between mid-February to the end of April, and *meher*, from mid-June to early September; the two seasons are also known in the literature as the short (or spring) season, and the main (or autumn) season. During years of good harvest, the average peasant practicing the bimodal system will have just enough food in hand from his *meher* harvest to tide him over to the *belg* season, and thus for many peasants in this system rural life means year-round labour and drudgery.

One may distinguish several agro-climatic regimes in Wollo, and briefly they are the following. For a more detailed discussion readers are referred to the studies published by FAO/UNDP (1984A: esp. Vol. 3 which contains some good maps); for the Ambassel-Qallu areas see maps prepared for the Red Cross studies of the same area (1986: Vol. II Maps).

Map 1. *Wollo: Zones and administrative division.*

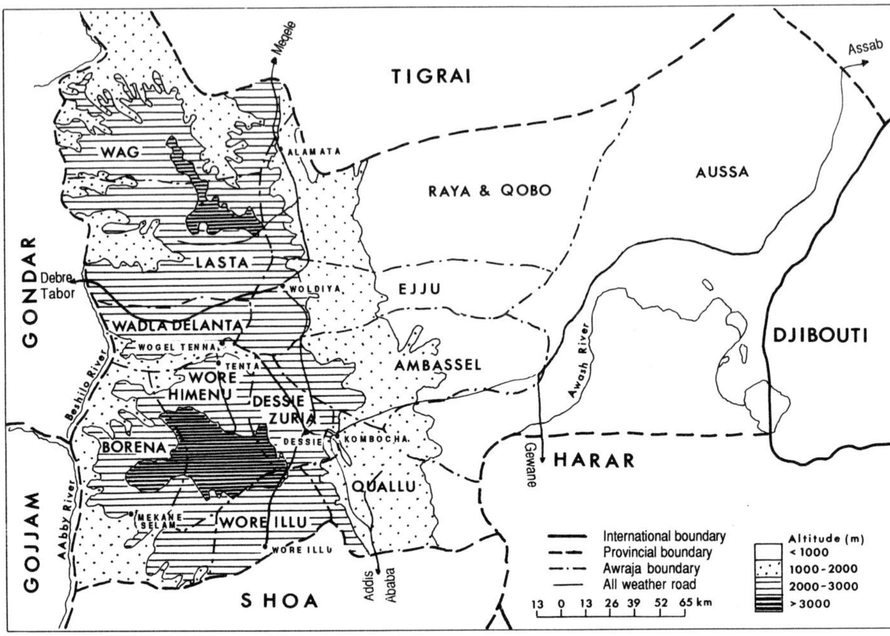

Eastern Wollo, the Afar lowlands. This area, most of which lies in Aussa awraja, has an altitude of 200 to <1,000 metres (m) above sea level. The mean temperature during the growing season is 20 to 28°C, and the mean annual rainfall is not more than 200 millimetres (mm). There is not enough precipitation during the belg season for crops, and even during the meher months the moisture is hardly sufficient for growing food. The length of the growing period (LGP)[1] in the main season is 45 to 75 days. The area is mostly given to nomadic pastoralism, although here and there small plots are cultivated by using simple hand tools.

Eastern Wollo, the foot of the escarpment. In Ambassel, the area stretches from near Bati in the east to the area along the Addis–Asmara road in the west, and to about 20 km south of Wichalae town in the north. The

[1] "A growing period is a part of the year during which the moisture supply from precipitation and soil water storage and the temperature are adequate for crop growth" (FAO/UNDP 1984A: Vol. I, P. 6).

elevation here rises from 1,000 to 1,800 m above sea level, and much of the region falls within the *qolla* (see below) ecological zone. The mean temperature during the growing season may stand at about 20°C, and the mean annual rainfall reaches 500 mm. The belg season has an LGP of 45 days, and in the main season LGP is between 120 and 150 days.

The escarpment itself begins from 1,800 m altitude, and at its higher elevations receives on the average 1,200 mm of rainfall a year, the highest in Wollo. Here—and this is also true for other parts of Wollo—frost becomes a hazard at altitudes above 2,000 m.

Western Wollo. This is mostly high plateau, a good part of which lies between 1,800 m and 2,800 m, although the areas with altitudes above 3,000 m are not insignificant. Here, at lower elevations, the mean annual temperature during the growing period measures 15 to 20°C, while at higher altitudes it reaches 10 to 15°C. The length of the growing period in the area as a whole in belg may be between 45 to 60 days. In the main season, LGP in the northwestern region (including Lasta) will often be 120–150 days, while in southwestern Wollo it is 150–180 days. In the high alpine zones, especially in Borena awraja, LGP may reach 200 days.

A brief discussion of the agro-ecological setting of our province is now in order. This subject, like many others in Ethiopian rural studies, is surrounded by some controversy, nevertheless, and Mesfin Wolde Mariam's structures (1985: Annex 2) notwithstanding, the effort to design, at least for analytical purposes, regions with common agricultural and environmental characteristics is a valid and useful exercise. There are at present two main zonation systems frequently employed by researchers, each based on different principles and methods.

The series of *Ethiopian Highlands Reclamation Study* (EHRS) commissioned for the Ministry of Agriculture by FAO (see MOA-FAO Working Papers in the Reference section) has adopted a zonal classification system based primarily on a region's supposed agricultural potential and crop specialization. This system has also been taken up by farm systems researchers working with ILCA (eg. Gryseels and Anderson 1983), and by some MOA experts. Briefly this approach divides the country into three major zones, called low potential cereal (LPC), high potential cereal (HPC), and high potential (or horticultural) (HPP) production zones. The main weakness of this exercise (and some of Mesfin's criticism of it is well taken) is, in our opinion, that the concept "high or low potential" is imprecise and subjective, one not based on accurate measurements or rigorous evaluation. There is no reason, for instance, why one should not have used three, instead of two, categories of potentiality, i.e. high, intermediate and low. In some EHRS reports Wollo falls into two zones, western Wollo being in the LPC and eastern Wollo in the HPC categories. Yet the main surplus awrajas and woredas of Wollo are in the west, while a great portion of eastern Wollo has often suffered from food shortages.

Westphal (1975: 83ff) has on the other hand attempted to popularize a zonal system based not on the alleged potential of a region but on the system of production in practice there. By this is meant three important elements: the chief purpose of production, the labour process involved, and the technology employed. Four major zones are identified using these criteria: the seed farming, the *ensat*-planting, the shifting cultivation, and the pastoral complexes. The merit of this system is that it looks at a region from the perspective of how production is organized rather than how good or bad the farming conditions are, and it takes into account the interplay between ecology and agriculture. Thus central and western Wollo fall within the seed-farming complex (which we shall discuss under the rubric of the peasant mode of production), while eastern Wollo is part of the pastoral complex.

If the reader is dissatisfied with both these approaches to zonation, there is a third system which is based on traditional peasant "climatology", which we shall refer to frequently in this work. The problem with this system, however, is that not everybody agrees where one zone ends and another begins, which makes the system less precise and open to charges of arbitrariness (see EHRS reports, Westphal: 39, Huffnagel: 78, for the different approaches).

Ethiopian peasants recognize the following five climatic zones, each of which is associated with particular (though not well-defined) agricultural regimes (the figures provided are my own and they are based on various technical maps). Some add a sixth zone called *kefil-bereha* (semi-desert). (See EMS, 1979: 2).

A. *Bereha:* This covers land below 600 m altitude, and as the name implies (lit. "desert") it applies to all areas where desert conditions obtain. The eastern lowlands of Wollo lie in this zone.

B. *Qolla:* Here the altitude ranges from 600 m to 1,600 m above sea level, with a mean annual rainfall of between 200 mm and 300 mm, and a growing period in the main season of 45 to 75 days. Parts of central Wollo, except those areas near to the escarpment, fall in this zone.

C. *Woyna-degga:* This covers areas between 1,600 m and 2,300 m elevation, with a mean temperature during the growing period of 20°C, annual rainfall of 500 mm (slightly higher in the upper reaches of the zone), and a *meher* LGP of 120 to 150 days.

D. *Degga:* Rising from 2,300 m to 3,200 m above sea level, lands in this zone (mostly high plateau) enjoy a temperature range of 15°C to 20°C, rainfall over 600 mm, and an autumn season LGP of over 150 days. Frost is a serious and frequent hazard especially in the higher reaches of the zone.

The main agricultural areas of the province fall within the *woyna-degga* and *degga* zones, which are also where the main concentration of the population is to be found. In the degga zone, lands in the lower altitudes have more agricultural potential than those in the higher elevations.

E. *Wi'rch* (or alpine zone): This refers to areas over 3,200 m altitude. Cold, wet, and wind-swept, settlements here receive rain fall of up to 1,000 mm, temperatures mostly below 15°C, and a growing period of over 180 days.

It is estimated that about 42 per cent of the land area of Wollo lies in the *qolla* and *bereha* zone, some 33 per cent in the *woyna-degga* zone, and about 27 per cent in the *degga* and alpine zones. The zonal breakdown of *woredas* in Ambassel *awraja* is given in Table 2.

Table 2. *Zonal breakdown of Ambassel (% of land area)*

Woreda	Qolla and below	Woyna-degga	Degga and above
Ambassel	20	33	47
Worebabo	60	37	3
Tehulederae	28	60	12

The evidence available in MOA records in Dessie suggest that the actual cropland area of the province is very small, most of the land being unsuitable for rainfed agriculture. About 45 per cent of Wollo is unusable, and this includes the arid Afar lowlands; another 25 per cent is mountain and slope, and land under rivers, gorges and still water. Only an estimated 20 per cent is crop and grazing land. The forest cover of the province has shrunk to a dangerously small size (less than 5 per cent), and the chief source of fuel for home use is now animal dung.

The picture in Ambassel is slightly different in that the crop and grazing land—about 40 per cent in our estimation—is proportionally much larger than in the province. However, here too, the forest cover has over the years been aggressively destroyed so that at present less than 2 per cent of the awraja is covered with forest.[1]

Wollo has a relatively large population, about 3.4 million rural inhabitants, and this, coupled with the limited agricultural potential, and the *ever-increasing* degradation of its resources has contributed greatly to its endemic travail. Excluding the arid lowlands (which are thinly inhabited by nomadic herders), the density of population of the province comes to about 225 persons per square kilometre, a high figure by Ethiopian standards.

As the pressure on the land has kept growing and land has become more and more difficult to obtain, an increasing number of peasants have taken to slope and mountain cultivation. In the *degga* areas of Ambassel and in many parts of western Wollo it is not just mountain slopes that are farmed but the summits of the mountains themselves. In the central areas of western and northwestern Wollo, something like a quarter of all farms are clinging to hillsides and perched on hill tops. One observes here and there whole mountains, stripped clean and brought under cultivation, and looking at times like a huge animal that has been skillfully skinned. In the rugged terrain west of Wichalae in Ambassel, on the higher altitudes of the escarpment, men farm tiny plots of land on the sides of steep mountains and precipices suspended on a rope.

Each year more and more peasants are banished to the mountains, and more and more (meaning really, poorer and poorer) mountain slopes are brought under cultivation. As a result there is now emerging in the province what may be called a *mountain peasantry* which will eventually become distinct and separate from the rest of the rural population. Since the state provides very little that is beneficial to the peasant but siphons off a good deal of the latter's harvest, a large number of these mountain peasants will soon be driven to break their ties with the outside world, and to retreat into their rugged fortresses to lead a life of independence in sublime isolation.

What flows from this complex agro-ecology will be discussed at some length in the course of this study, but let us say a few words about the

[1] For a different estimation of land use and land cover in Wollo and Ambassel, based on aerial maps, see FAO/UNDP (1984B: 44, 74).

interplay of environment and production.

As a general rule, the size of holdings becomes smaller as one moves from the lower to the higher altitudes, and the level of income of peasants likewise decreases. In the lower plains of southwest Wollo, average holdings measure as much as 2.0 hectares per household, while in the *degga* areas of Ambassel the figure is between 0.5 and 0.75 hectares. Similarly, as holdings become smaller, and as men are driven to work steeper and steeper slopes, the plough is replaced with the hand tool and traction power elevations have less grazing potential than those lower down, and this has a bearing on the system of peasant livestock management.

The crop, and to a certain extent livestock specialization that ecological divergences give rise to, is one of the key elements that sustains the economic ties between highlander and lowlander. Without this, the two regions would have no need of each other, and the existing relations of exchange would long have been replaced with those fuelling hostility and ill intentions. Table 3 shows the major crops grown in the main agricultural zones.

Table 3. *Main crops in the three zones of Wollo*

Degga	Woyna-degga	Qolla
Barley	Sorghum	Sorghum
Teff	Maize	Maize
Horsebeans, Peas	Teff	Sesame Seed
Lentils	Wheat	Cotton
Emmer Wheat	Noug (niger seed)	Millet[a]
Linseed	Beans and Peas	

[a] This crop is not widely grown in Wollo

A third element in the interplay between environment and economy has to do with the relative importance of spring agriculture and the system of resource management employed by peasants in the various zones. As noted earlier, the *belg* season assumes greater importance in the higher altitudes of the *woyna-degga* zone and in all zones above that. We do not know precisely how significant *belg* cultivation is for the Wollo peasantry. According to one source, *belg* agriculture makes up 40 per cent of the annual agricultural production of the province (NEERNDRC: 32), but this in our opinion is an exaggeration; the area given to *belg* crops is smaller, and the yield obtained in the season is often lower than that in the main season. According to CSO, *belg* crops on the average cover about 12 per cent of the total cultivated land in the province and *belg* production is about 12.5 per cent of the main harvest. This too is an underestimation, and the real facts probably lie somewhere in between the two estimations. A Ministry of Agriculture survey conducted on the eve of the famine (MOA 1984B: 53) reports that 63 per cent of Wollo peasants never farm their lands in the *belg* season, 10 per cent do so on a regular basis, and 27 per cent only occasionally. According to the records of MOA in Ambassel, close to one-third of the peasants in the awraja regularly practice *belg* season agriculture, and the failure of the rains or of crops in the season will put severe strains on such peasants. On the other hand, the year-round cultivation of the land—which is what bimodal agriculture involves—leads to rapid soil exhaustion, and to mitigate this and to maintain soil fertility peasants employ a fairly complicated crop rotation system.

In these circumstances, resource, and specially food management assume great importance, second only to farm management. The peasant has to balance the needs of his land with the needs of his family, and the decision to select a particular crop mix for one season is arrived at after careful consideration of all his options. Cossins (1974: 42) argues that there is no long-range planning for crop rotation, and the decision to rotate is based on a year to year basis. This is not a flaw in traditional agriculture since the factors that influence a peasant's decision, such as the weather, the condition of the land, the yield obtained the previous season, the needs of his family and the claims of outside forces change from year to year, and sometimes from season to season.

Finally, environmental diversity must also be viewed in relation to the system of livestock management and ownership adopted by peasants in the various ecologies. Those in the arid lowlands can only specialize in animal husbandry as agriculture (rainfed agriculture, that is) is out of the question for them. On the other hand, there are differences in livestock management and ownership among sedentary agriculturalists arising from micro-environmental differences. Cossins (ibid: 11ff) has shown that in the higher altitudes sheep become more important than cattle, while lower down, specially in the areas below 2,400 m, cattle take precedence over small stock, and further below, in the *qolla* areas below 1,500 m altitude, goats assume importance. Moreover, livestock mortality is higher in the *degga* and alpine zones, and decreases as one descends into lower altitudes. All these factors influence the pattern of livestock disposal and ownership followed by individual peasants, and affect the nature of the overall livestock market.

A few words must be said about the degradation of resources in Wollo. By resource degradation is meant the permanent loss of productivity of capital assets or the complete loss of the assets themselves. Recent studies have begun to reveal the magnitude of the resource loss of the country, although micro- or case studies of different regions showing the severity of the problem are not well-advanced (see MOA-FAO 1984B: Ch. 6 for a bibliography on the subject, MOA-FAO 1985: Annex 3). Specialists now estimate that nearly 2 billion tons of soil are being washed away every year, most of it from the highland regions, particularly from Wollo, Gondar, Tigrai, Gojjam and Shoa. Of this, some 10 per cent is irretrievably lost because it is carried out of the country by highland rivers. The rest is redeposited as sediment elsewhere in the country, usually in areas which are of little use agriculturally. Because of the loss of valuable top soil, many highland soils have a low nitrogen content. This explains the importance attached to nitrogenous plant cultivation in Wollo where on the average 20 per cent of the cultivated area is given over to such crops.

In the areas specializing in grain cultivation, particularly in the traditional famine areas, an average of about 100 tons of soil per hectare per year is permanently lost. The highest rate of erosion within the seed-farming regions occurs in Wollo, with the higher elevations of the province contributing most to the destruction of ecology compared with the lower. In about two and a half decades, resource degradation could wipe out about 15 per cent to 20 per cent of the province's farmland affecting the livelihood of well over half a million peasants.

The chief cause of this massive loss of resources is human activity, fully aided by nature. Poor, or rather abusive land use practices, slope cultivation, deforestation and overgrazing are the chief causes of soil erosion, and more than half of the annual losses suffered by the Wollo peasantry is set off by rainfall. A good percentage of peasants in the province are well aware of the problem of resource degradation, and of the connection between it on the one hand, and drought and famine on the other (Yeraswork and Solomon 1985: 18–22); but their own conservation practices are ineffective (the increasing severity of the problem has made them more and more so), and the resource enhancement programmes promoted by government and international donor agencies are limited in their impact and scope.

4. The Economy of Wollo

Noel Cossins, one of the most discerning students of rural Ethiopia, made this remark about Wollo (ibid.: Zone 2, p. 10):

> Wollo is virtually impossible. There is such an obscuring weight of disbelief, suspected innuendo and antagonisms; such a mess of mis-government at petty levels, and such a lading of landlords that there is almost nothing to start with and nowhere to start that will not go wrong or sour. Changes and innovation will inevitably be resisted and resented, and even if a foothold of development thought and action is established, its small flower would soon be checked by the smothering welter of the weeds of an entrenched and stagnant society.

This was written on the eve of the revolution, and of course the landlord has since disappeared, and some radical, structural and institutional changes have taken place in rural Wollo. Wollo is a fragile society now, made even more fragile by the continuing price it has to pay for the radical reforms themselves, and by increased poverty and increased hardship. Nevertheless, there is something in what Cossins has to say: "Wollo is still a stagnant society, and what little development activity has been going on in the decade since has barely touched the peasant."

Even by Ethiopian standards Wollo is one of the least developed provinces in the country. Land transport has not expanded much since the mid-1960s, and as much as two-thirds of the peasantry is inaccessible by road. Western Wollo is far more poorly served in this respect than the east. Communication services for much of the province are poor and outdated. The province is among the least urbanized in the country, just ahead of Illubabor and Wollega provinces in this respect. Dessie, the capital, with a population of about 69,000 is the biggest town, followed by Kombolcha nearby and Woldiya to the north with 16,000 each; there are no other towns with more than 15,000 inhabitants. In Ambassel, Haiq (pop.: 5,000) is the capital and the largest town.

Wollo peasants are among the poorest in the country. A recent survey found that peasants here had the lowest standard of living compared to peasants in ten of the country's regions—below the standard of peasants in Bale and Gamo Goffa provinces which are generally considered to be severely underdeveloped (MOA-FAO 1983: 28–30). The study also indicates that Wollo peasants are far less integrated into the market economy than the other peasants. While education and mass literacy has expanded considerably in the last decade, the latter thanks to the post-revolution literacy campaign, very little progress has been made in the areas of health and welfare, and virtually none in those of employment and basic infrastructure.

While it is true that Wollo has been badly neglected insofar as development activity is concerned, this neglect has not been the product of political ill-will as has been alleged by some with reference to the Old Regime (Shepherd 1975: Ch. II), nor has it stemmed from some other antipathy to this area on the part of planners and decision-makers. There are several serious constraints to development which have not often been taken into account.

First and foremost, the rugged topography of the area discourages investment on heavy infrastructure, especially since the returns on such investments are known to be limited. There are no untapped resource-rich regions to open up, no Eldorados to discover, no new and promising economic activities to stimulate. Wollo has not been a food surplus area—not in the same league as Arssi or Gojjam anyway—and except for an enclave in the Afar lowlands and the Awash delta (both of which are neither historic-culturally nor ecologically part of Wollo) the province does not produce cash crops for the world market. There are, besides, no known mineral, energy or other economic resources to wet the appetite of domestic or foreign investors. Wollo in brief is a land of humble potentials densely inhabited by millions of small and hardy peasants who have for centuries practiced what they know best, namely subsistence agriculture.

In this chapter we shall be concerned with the broad outlines of Wollo's economy, and particularly as it relates to employment. Any crisis that occurs in the countryside triggers a heightened demand for short-term employment by the peasantry which often flocks to the urban areas in search of jobs and income earning opportunities. How much support can the economy provide the rural population? The information available in this area is extremely patchy, and the reader should not expect a detailed analysis. Let us first begin with the urban sector.

It will not be an exaggeration to say that if the urban economy of Wollo was to cease functioning completely, the rural population would not be seriously inconvenienced. Except for trade in agricultural goods, which the peasant supplies to the urban dweller, there is very little economic interaction between town and country. Of course, the towns absorb some of the rural unemployed; since certain services, and all government offices are located in the urban areas, the peasant is forced to make an appearance there periodically. Some peasants may also have friends or relatives living in towns and may from time to time visit them there. Be that as it may, the two sectors, urban and rural, are not, in the deeper economic sense of the term, mutually interdependent; indeed, the towns have been for decades in the past, and will continue for many years in the future to be, parasitically dependent on the rural economy.

A visit to a rural market is instructive in this respect; revealing clearly how the rural and the urban lead a separate and autonomous existence. We visited several woreda markets in Ambassel, and more than once the markets in Haiq, Golbo and Wichalae towns. Roughly 80 per cent of the products on sale in these markets (by kind) were of rural origin; of the remaining products originating in urban areas, about half were dispensable, i.e. the peasant could either do without them if the need arose, or could easily find satisfactory substitutes. The variety of manufactured goods available in rural markets was much greater before the revolution than at present, so too the level of integration of the peasant into the market economy (cf. Baker 1986: Ch. 10).

There are three main sources of employment in the urban economy, and all three have not been in a healthy condition for many years for reasons having to do with obsolete equipment and low productivity in the case of manufacturing industry, and unfavourable government policy, and a crippling state bureaucracy in the case of the other two. The three sub-sectors are urban industry, small-scale and handicraft production, and petty trade.

Wollo has only four manufacturing enterprises—the fourth, a new textile mill in Kombolcha started operations at the time of our field work. The three older establishments, namely a meat packing and leather-making plant in Kombolcha, and a soft drinks bottling firm in Dessie, employ a total of about 1,800 workers, and when the textile factory becomes fully operational the industrial working force may rise to 4,200. According to the government's Ten-Year Plan, industrial employment in the province is expected to reach a little over 7,000 at the end of the plan period, i.e. 1993/94. Kombolcha, where three of the four industries are located, has good communication services, is the hub of the northeast–south road network and possesses sufficient power supply, and is thus well-placed to become the industrial and commercial centre of Wollo. However, the three older plants are operating poorly, as the equipment they are using is obsolete, spare-parts are difficult to obtain, and breakages are not infrequent (NEEPR Part 2:81–120). Needless to say labour productivity has been declining for many years. As things stand now industry has only a small role to play in stemming the tide of urban unemployment, and virtually none is helping mitigate the demand for jobs by the rural population.

The measurement of unemployment in Wollo (or elsewhere in the country) is virtually an impossible task. Unemployment figures available with the Ministry of Labour show only the tip of the iceberg, as it were; they include only those persons who appear at the offices of the Ministry and registered as unemployed. For example, job-seekers registered with the branch offices of the Ministry in the Wollo province were reported to number 5,840 in 1984/85; this must be an insignificant percentage of the province's total urban unemployed. It may be closer to reality to estimate that at least 20 per cent of the urban labour force in the province is unemployed; this figure is the same as the national unemployment estimate suggested by an ILO study mission a few yeas ago (ILO 1982: Part I, 16). An indication of the ability of the urban sector to provide employment may be gauged by the rate of successful placements of registered job seekers in the province. According to the Planning Office in the region only 10 per cent to 12 per cent of the registered unemployed were placed in jobs in the years 1984 and 1985. (NEEPR: Part 1, 30ff).

The other sources of employment in the modern sector (excluding government bureaucracies) are shown in Table 4.

There is no active handicraft industry in Wollo, and the little that is to be found is in the rural areas (and will be discussed further on). According to published HASIDA reports Wollo is among the least developed in the country as far as small-scale industry is concerned, just a little better than Gamo Goffa and Illubabor provinces where there are no small-scale industries known to HASIDA (HASIDA 1979, HASIDA 1985). Unpublished records from the HASIDA branch office in Dessie show that the main types of small-scale industries found in the province are grain mills (about 85 per cent), knitting outfits, and woodworks.

Table 4. *Employment in the modern sector (Wollo and Assab[a])*

Sub-Sector	No Employed
Agriculture	1140
State farms	1871
Electricity, energy, water	348
Construction	859
Transport & communication	1826
Distribution (commerce)	295
Banking and finance	835
Total	7174

Note: [a] The Northeast planning region consists of Wollo and Assab.
Source: NEEPR: Part 1, 39.

According to the same records 740 persons were employed by 201 registered small-scale enterprises throughout Wollo in 1986; on the average each major enterprise employs four persons per unit, although knitting enterprises and confectionaries are often family-owned and family-operated.

The real size of small-scale industry in Wollo is probably 3 to 4 times the number shown in HASIDA records in Dessie, since most enterprise owners do not bother to register with the agency for fear of being harassed by the state bureaucracy. Nevertheless, cottage industry and small-scale enterprise, in brief the petty mode of production as a whole is at present an insignificant force as far as providing employment or creating new economic opportunities are concerned. Needless to say, encouraging and well-intentioned policy reforms regarding the sector, and well-designed incentives for those active in it could change that considerably.

The third main source of income in the urban sector is petty trade and servicing, including in the latter category hotels, restaurants and drinking places. An attempt to estimate the contribution of this sub-sector to employment and income-generation is a perilous exercise, since no census or limited survey has been made about it either nationally or in Wollo. The shops, stores and other related enterprises registered with the authorities in the province will probably be outnumbered 2 to 1 by the unregistered ones, which deal with anything from distributing dairy products to selling home-made brews, and which are often carried on in the home. Suffice it to say that enterprises in petty trade are almost exclusively family-run ventures, and together with services provide a source of livelihood for more people than the two other sub-sectors combined.

Interestingly enough the very structure, and smallness of enterprises in this sub-sector make it easy for new entrants into the field. Petty trade and servicing is far more flexible, and requires far less capital than the other two. The peasant who comes to town to sell chickens and eggs door to door becomes a petty trader while he is doing that; the youngster selling confectionery after school hours requires hardly any capital.

Again, this sub-sector could play a more active role in the local economy if it were not for restrictive legislation, and inefficient distribution practices.

The Agricultural Sector: At present peasant agriculture predominates over all other forms of agriculture as shown below:

	% holdings	
	Wollo	Ambassel
Peasant farms	96.2	93
Cooperatives	1.2	7
State farms	2.5	Nil

Source: MOA records (Dessie and Haiq); NEEPR.

We shall discuss peasant agriculture at greater length in the next chapter.

Agricultural producers cooperatives (PCs) are most wide-spread in Ambassel and southern Qallu awraja. Table 5 contains some basic data about PCs in Ambassel.

Table 5. *Cooperatives in Ambassel awraja*

Woreda	No. of PCs	Membership			Holdings (Hectares)	No. Oxen (single)
		M	F	Total		
Ambassel	5	1342	142	1484	989	1227
Worebabo	3	838	56	894	739	622
Tehulederae	11	2058	78	2136	1372	1443
Total	19	4238	276	4514	3100	3292

Source: Records of MOA, Ambassel awraja, Haiq (October 1986).

We do not know how these cooperatives were organized but the data indicates that each member on the average brought with him/her into the organization 0.68 hectares of land, and 0.73 oxen; this latter figure may be misleading as the PCs must have lost some oxen during the famine. Let us assume that of the male members (we will not consider women for hard work for the time being) about 4,000 are deployed in farm work and the rest in other activities. Each male member will thus work only 0.78 ha of land, and similarly each pair of oxen will be able to plough 1.9 ha during the farming season; this assumes that all the PCs' possessions are cultivated during a season which is not often the case.

CSO reports (1984: 219) indicate that yield in cooperative farms in Wollo has declined progressively from 1979/80 (the year when data on cooperatives began to be collected) when it was 10.5 quintals per hectare (q/ha) to 1982/83 (the year before the famine) when it was 8.24. MOA records in Haiq suggest that yield for local PCs has stood at 12.3 q/ha for the last few good years. Taking this figure, and assuming that all of the local PCs' possessions were farmed we arrive at a total crop production figure of 38,130 quintals, or an income in kind of 8.4 quintals per member per season.

Some ILCA studies (Gryseels 1983) have shown that a highland peasant owning one ox can effectively work 1.9 ha of land, and another with a pair of oxen 2.7 ha. Obviously labour is poorly deployed and work inefficiently organized in PCs; and moreover, if our estimation of output and income for Ambassel PCs is accurate the individual peasant farmer is much better off than his neighbour who has joined a cooperative.

The reader will bear with us if we digress here to make a few remarks about cooperativization and agrarian socialism.

By 1979 the government was fully convinced that cooperatives were an essential element of rural transformation, complementing state farms which had been organized some four years earlier. The cooperative directive of that year, and subsequent legislation, required peasants to pool all their resources together and carry out production in association. Although, on paper peasants were to enter into cooperatives voluntarily, and although cooperativization was to be carried out gradually over many years, what happened in practice diverged considerably from both the letter and the spirit of the legislation. According to the Ten-Year Plan drafted in 1984, about half of all rural production, and 52 per cent of the cultivated area of the country was to be the cooperative sector by the end of the Plan period, an ambitious undertaking considering that the comparable figure in both cases in 1984 was a little over 1 per cent. The pace of cooperativization has accelerated considerably since 1985.

The basic assumptions on which socialist agriculture are based—i.e. collectivization as opposed to individual enterprise, association as opposed to isolated labour, large-scale as opposed to small-scale cultivation—are too well-known and need not detain us here. Briefly stated, the emphasis of proponents of agrarian socialism rests on the *organization of production*, while the decisive element in our opinion should be the *organization of the labour process*, involving as a condition the transformation first of the division of labour and the productive forces. The pooling of resources by peasants, and the cultivation of land in concert using traditional know-how and equipment—which is what cooperativization means in Ethiopia—is not an advance on traditional agriculture; since, in the end, each peasant will do what he would have done if he was working privately, no net improvement in output will be achieved.

Moreover, given the level of the productive forces of peasant agriculture, individual household production is rational and efficient judged on its own terms. The limitations of this system can only be overcome by the introduction of improved techniques, and by changes in the labour process. What about "association" and mobilization as a force of production? It may sound paradoxical but individual household production involves far more "association" than group labour organized under "socialist" programmes.

Let us return to our main subject. There are only two state farms in Wollo—Cheffa and Tendaho—but both have suffered as a result of poor planning and poor management. The Cheffa farm just south of Kombolcha has been operating in the red for many years, and officials in the area believe that it will continue to perform badly for many years to come. This may be part of the reason why plans are afoot to phase out all farming operations and to turn the farm into a fullfledged dairy enterprise. Tendaho, in the lower Awash valley, once a thriving enterprise, is now a pale shadow of its former self, and while it is not about to be closed down or changed into something else, it has ceased to be a source of seasonal employment for the highland peasantry as in the days before the revolution (NEEPR: Part 1, 37).

An examination of the available evidence shows clearly that rural production in Wollo has been in stagnation for more than a decade. While drought and famine have played a part, this poor record is also a result of the following factors: miniscule holdings and land fragmentation—and both discouraging the widespread use of modern inputs and the dissemination of innovations; the increasing degradation of the soil and other rural resources; disincentives to peasants stemming from unfavourable agricultural legislation and policies; and destabilization of peasant agriculture brought on by periodic rural conflict, and frequent changes in rural policy.

The data in Tables 6 and 7 is for the period 1974/75 to 1978/79, and 1979/80 to 1983/84. Unfortunately, the information comes from two different sources, and it is hence difficult to compare one table with the other. However, the evidence does show that neither gross production nor farm productivity improved significantly in the two periods covered by the data.

Table 6. *Area, production and yield of major crops in Wollo 1974/75–1978/79*

	1974/75	75/76	76/77	77/78	78/79
Area[a]	223.1	212.2	205.8	211.7	201.5
Production[b]	2297.4	2142.5	2166.3	2226.6	2031.3
Yield[c]	10.3	10.1	10.5	10.5	10.1

Note: [a] Thousand hectares [b] Thousand quintals [c] Quintals per hectare.
Source: MOA-CSO 1979.

Table 7. *Area, production and yield of major crops in Wollo 1979/80–1983/84*

	1979/80	80/81	81/82	82/83	83/84
Area	552.7	631.3	693.7	601.4	490.7
Production	7654.0	10 164.6	8709.8	7926.0	5068.5
Yield	13.9	16.1	12.6	13.2	10.3

Note: Same as note in Table 6.
Source: CSO 1984.

Wollo's prospects in agriculture are quite discouraging, but the area has one resource which, if properly developed, could become an important source of income for the rural population. We are referring to the livestock potential of the region. For the peasant here livestock is not just a form of savings but the most important means of earning cash. Table 8 contains a livestock "census" for the province as a whole and for Ambassel awraja; the data was obtained from MOA, Wollo, and covers the period between 1982/83 and 1984/85. The figures are actually estimates, and those for 1984/85 indicate the loss of animals suffered by peasants in the area due to the famine of that year.

Table 8. *Livestock population of Wollo and Ambassel 1982/83–1984/85 (Thousand)*

	Wollo			Ambassel		
	1982/83	1983/84	1984/85	1982/83	1983/84	1984/85
Cattle	2222	2384	1441	160	236	142
Sheep	2650	2650	1660	200	200	110
Goats	1960	1960	1472	150	150	105
Equines	990	990	822	65	65	48

Source: MOA, Livestock and Fishery Dept., Dessie, 1986.

The livestock business in Wollo is an active one, with several important large markets, of which Korem, Haiq, Dessie, and two other centres in western Wollo are the most active, and numerous feeder markets scattered along the major livestock routes throughout the province (see Map 2). Markets in the first category are mainly involved in the "export" of livestock to the two large consumption centres, i.e. Addis Ababa and Asmara; the first is a very important market for sheep, the latter for cattle and goats. Peasants here are willing sellers of livestock, particularly sheep and goats, and if the price is right a peasant will deplete his flock to take advantage of the market. A study prepared in 1975, probably the only one of its kind, suggests that of the total number of sheep and cattle that arrive in Addis and Asmara every year from the highlands, the largest number originate in Wollo (LMB 1975: 11–16). If on top of this one includes the large number of livestock marketed in the lowlands by the Afar (real magnitude unknown), one will have a good enough picture of the potential of livestock as a development resource for the province.

However, peasants do not often obtain the full benefits of their resource, for a variety of reasons. Mortality among animals, specially

Map 2. *Ambassel: Livestock flows during famine.*

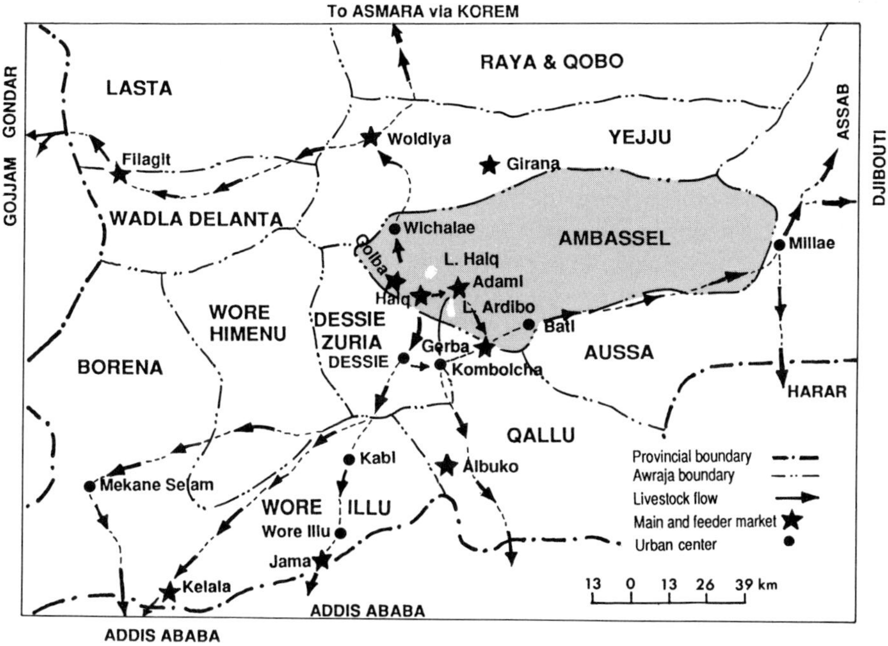

sheep, is high, flock maintenance is a problem as grazing areas are limited, and subsistence-based peasants are always in need of cash to meet their obligations and to purchase food and other necessities. Predators are also a problem, although loss of animals, especially of small stock due to them is not high enough to be a major concern. For these and related reasons, animals are sold young or before they reach their full potential weight, i.e. when their price is low. The most serious hazard to livestock development, however, continues to be drought and famine, when large numbers of animals die or are disposed of at rock bottom prices. If the MOA estimate shown in Table 8 is accurate, Wollo may have lost, through death or crisis-induced disposal, as much as 30 per cent of its livestock population in the last famine.

While the province as a whole is not a food surplus area, there are localities, woredas and awrajas which produce enough to feed themselves or a surplus to sell to others. The deficit areas are supplied grain by the surplus areas and peasants in the latter use the money thus acquired to buy livestock (specially cattle) from those in the former.

This interchange between deficit and surplus zones is greatly heightened during times of stress, and will continue to operate in this manner until the means of exchange of one or both zones is fully exhausted. Of the twelve awrajas of Wollo only two may be characterized as food surplus areas. According to MOA (Dessie) records and oral interviews of WPA Leaders the three major food producing zones are the following (awrajas):

Crop deficit areas: Wag, Lasta, most parts of Raya Qobo, eastern Qallu, eastern Ambassel, and Aussa.

Self-supporting areas: Wore-Himenu, Wadla Delanta, Yejju (The southwest corner of this awraja may occasionally produce a surplus), western Ambassel, southwestern Qallu. These areas quickly become deficit areas if adverse conditions arise.

Grain-surplus areas: Borena, Wore-Illu, western Dessie, Zuria.

When famine occurs in the northeast it is rarely the case that all areas are affected equally and at the same time. This is also true of Wollo, and is partly a result of the province's diverse ecology and climate which allows some localities to escape disaster while others are laid waste by it. This means—and if this is what Sen (1981) is arguing it is hardly original—that it is not the total unavailability of food in the area that the hungry are suffering from but the lack of resources or power to acquire food, or the lack of knowledge of its availability. In the 1984/85 famine the grain surplus areas noted above were only moderately affected, and this did not seriously reduce their production; they thus continued to act as suppliers of food grain to peasants in other parts of the province during a good part of the crisis.

It may be worth noting that the grain surplus areas are those where average holdings are larger, where much of the farm land is bottom land and relatively less slope cultivation takes place, and where prolonged drought, or periodic rain failures are not as frequent as elsewhere.

5. The Peasant Mode of Production

Studies of rural production in Ethiopia have often been conducted outside general theoretical frameworks, and the result has been a plethora of empirical-technical reports and documents, and a dearth of theoretical and conceptual-oriented works. It is enough to look at the EHRS monographs on highland agriculture (MOA-FAO series, see also FAO/UNDP 1984 series), and the ILCA-sponsored studies of farm systems (Gryseels and Anderson 1983, Anderson 1985) to realize how much the empirical predominates, and how far the theoretical approach has been shunned. This is not to say that these and similar studies are all poor, on the contrary some of the works produced of late, particularly the case studies, are of high quality, and more of them are needed in the future. We are only pointing to a neglected field, and suggesting at the same time that researchers, specially academic researchers, should attempt to redress the balance.

A detailed theoretical analysis of peasant production is outside the scope of this study, but we shall hold that Ethiopia highland sedentary agriculture is best understood as a mode or system of production with distinct characteristics of its own. A distinction should be made between sedentary agriculture and pastoral production, or swidden cultivation which are themselves distinct production systems involving populations living in the western, eastern and southern periphery of the country. While there may be variations in one or more aspects in one region or another, arising obviously from differences in environment, land use, past practices, the dominant parametres of the system remain essentially the same everywhere. The specificities of the peasant mode of production that are important for our purposes are the following:

(a) The family is the unit of labour, of production and consumption. The chief aim of production is consumption and not accumulation, and wage labour is not an element in production. The family as the unit of labour means more hands are valued, and this leads to high rates of demographic expansion. Social stratification is greatly influenced by demographic cycles. In post-revolution Ethiopia, poor peasants are often those at the beginning of the demographic cycle (young households) and at the end (old families). In addition, livestock and disposable assets determine social status, and land is no longer an important factor.

(b) The market plays a supplementary role in peasant self-maintenance. The rural market is of course an important vehicle for household resource management, but it does not determine, in the last instance, a peasant's land-use practices or cropping plans. The demands of outside forces—government, creditors, etc.—influence peasant decisions far more than market behaviour. The segmentary nature of the rural economy, the limits to peasant mobility, and low disposable surplus help to make the market a secondary force in rural production. To this must be added what may be termed the ethic of peasant consumption: very often there is a socially accepted level of income, and peasants tend (at least in traditional northern Ethiopia) to be less motivated to labour once this level is attained.

(c) The stock of technical knowledge employed in production is the same for all households. Peasant know-how, productive practice and technology is, ecologically speaking, highly destructive. The peasant, on the other hand, is helpless in the face of long-term environmental degradation. Moreover, this technical parity among peasants rules out competition and this in turn discourages innovative endeavour. The chances for the diffusion of innovation is greater where land is not a limiting factor in production, and where success is adequately rewarded. The process however is long and hard.

(d) Periodic food crisis is inevitable in the peasant mode of production (just as periodic economic crisis is said to be unavoidable under capitalism). High rates of surplus extraction by outside forces, population pressure, and environmental stress have a damaging effect on the system which cannot easily adapt to such pressure. Food crisis need not however lead to famine and large-scale deaths.

In the pages that follow we shall discuss some of the essential elements of peasant production in Ambassel awraja. Peasants in the awraja are organized in 85 peasant associations (PAs), and each household must register all its members with the PA of its area. How much land a household received during land apportionment was based on the number of individuals in it, and the same was true with regard to emergency relief assistance. In theory, therefore, PA rosters should show, at least in figures, all the people living in the rural communities, and should serve as a reliable though crude census of the rural population. This is not the case however in practice as Table 9 indicates.

Table 9. *PAs and members in Ambassel by Woreda*

Woreda	PAs	Hhd Head Membership			Total Membership[a]			1984[b] Census
		M	F	Total	M	F	T	
Ambassel	34	18 468	3725 (17)	22 193	49 803	48 616	98 419	119 354
Worebabo	23	15 826	4577 (22)	20 403	44 595	41 827	86 422	85 302
Tehulederae	28	16 997	3533 (17)	20 530	44 079	43 567	87 646	152 508
Total	85	51 291	11 835	63 126	138 477	134 010	272 487	365 645

Note: Hhd = Household. Figures in brackets show the percentage of women members.
[a] Includes all Hhd heads and all household members.
[b] Rural population only.
Source: Records of each Woreda PA, October 1986. The PA rosters were prepared in early 1985. For census, CSO 1985.

The difference between the census and the PA figures is not insignificant especially for Ambassel and Tehulederae woredas, and the only explanation is that some peasants are not registered with their PAs. This was confirmed for us by the chief extension agent of Ambassel woreda who stated that a good number of the mountain peasantry of the woreda, i.e. those living in the *degga* and alpine zones, have refused to register with their PAs and to participate in PA or MOA initiated activities. This kind of peasant behaviour, according to the agent, became pronounced after the land distribution of 1983. In the third woreda, the discrepancy between the PA and census figures may be due to the fact that some of the peasants in the lowland areas lead a seminomadic life and may often not bother to register. We should note in passing that female-headed households in the awraja make up about 19 per cent of all households.

As was noted earlier, land redistribution was carried out in the awraja only recently, and, according to extension agents in the area, the major reason for it was twofold: to try to equalize holdings as much as possible, and to clip the wings of largeholders ("rich farmers" some called them:

Ye-merrait kebbertae) who, these agents believed, were evolving into a new class and a "reactionary force" in rural society. The formula used to allocate land to families was a relatively simple one: each PA allotted an equal share of land from its land fund (excluding grazing land) to *each family member* registered in the organization. It was the responsibility of the head of the household to have all his family members registered in the PA. Most PAs allotted three grades of land in this manner, each family member getting an equal amount of land in each grade; the exceptions were some PAs which divided the land in their possession into only two grades. Land grading is a tricky business, and what may be taken as high grade in one locality may be considered middle grade in another. Home plots were not as a general rule included in the redistribution.

How much land a family obtained depended on the size of the family, and on the land fund at the disposal of the PA. A PA with a small disposable land area allocated a small unit to each individual, and vice-versa. In Table 10 are four examples from four different parts of Ambassel (two from Ambassel and one each from Worebabo and Tehulederae woredas) showing the actual size of plots distributed to each individual in the respective areas.

Table 10. *Plot sizes distributed to household members*

Examples	Graded plots per person (sq.m)			Plot size Hhd of 5
	Fertile	Medium	Poor	Hectares
1. Degga-A	400	625	900	0.96
2. Woyna-degga-A	900	1200	Nil	1.1
3. Qolla-W	900	1200	2500	2.3
4. Qolla-T	625	1225	2500	2.2

Note: Sq.m = square metres. A = Ambassel, W = Worebabo T = Tehulederae woredas.
Source: Oral interviews of PA leaders.

Thus a household of five persons will have acquired from 0.96 to 2.3 ha of different grades of land depending on the area and available farm land of its PA.

There are three important points that should be noted here and that is a direct outcome of the land distribution. (a) For most peasants, especially those above the *qolla* zone, the family farm measures between

1 and 1.5 ha; in the alpine zone, plot sizes may be between 0.5 and 1.0 ha. On both cases, there is barely enough land for basic sustenance. (b) In order to rule out unfair treatment of families, land distributors were forced to maintain and in some cases exacerbate land fragmentation which has long been a major problem in the area. (c) Land-redistribution (in this case land reform as a whole) has not further undermined the status of women; each married woman is entitled to a plot of land equal to that of her husband which she can withdraw from the family's possessions if the husband divorces her. Moreover, reform has not made the rural family fragile, or more susceptible to breakups than formerly as has been alleged by some. This is an important point and we should say a few more words on it.

McCann (1986 Ms.) has argued that divorces have become more frequent in the Ankober area of northern Shoa (an area with similar characteristics and farming practices as Wollo), and food crisis tends to exacerbate the fragility of rural households. When breakups occur it is usually women and the weaker members of the family that suffer most. We believe McCann's arguments are based on hasty observation, and faulty information.

The inclusion of women in land distribution has made them an important force in the family, and a source of not just additional labour, but additional land to the family on their own right. Frequent family breakups will involve frequent breakups of family possessions which will give rise to uncertainty in production and insecurity in tenure—evils most peasants try hard to avoid. Moreover—and this point has rarely been considered in Ethiopian rural studies—peasant women are capable of leading a life independent of men (witness the large number of female-headed households in Table 9 above), whereas no male peasant will be able to cope, even for a few weeks, without the support of a female partner. The immense labour and know-how involved in the processing, preservation and preparation of food alone—exclusively a woman's domain—will impose severe limitations on a peasant's ability or propensity to change his wives at will. Finally, community values discourage anyone from indulging in frequent marriages and divorces, and besides, no self-respecting woman will agree to marry or live with a man with a record of unstable family relations.

Our survey did not include questions related to this subject, but there is indirect factual evidence to support our arguments. In the field enumeration carried out for the Food-for-Work Programme in September 1984 (a high point in the famine) by Yeraswork and Solomon in north Shoa (see report 1985) peasants were asked if they had recently divorced or remarried. A great majority answered in the negative, and stated further that they have been with the same spouses for many years. The survey also found that divorce rates were not higher here than in communities elsewhere in the country.[1] Secondly, a survey of Ambassel and Qallu awrajas carried out by the Ethiopian Red Cross in February–March 1986 found that all the male respondents in the sampled population (i.e. 85 per cent of the household heads interviewed) had wives at the time of the interview. The report does not say anything about the 15 per cent female household heads in the sample, but points out that 94.5 per cent of the males were monogamous, 5.3 per cent had two wives each, and the rest three or more wives. It should be noted that Qallu awraja was one of the worst hit areas in Wollo (ERCS 1986, Vol. III:84).

The basic source of draught power is of course the ox, but in Ambassel (as in Wollo as a whole) this critical resource is not evenly distributed. Reliable data on oxen possession was not available to us at the time, but awraja extension in Haiq agreed that the following estimation was a fairly realistic one up to 1983 i.e. before the onset of the drought in the area: about 5 per cent of households in the awraja with more than two oxen each; 35 per cent of households with two oxen; 40 per cent with one ox; and 20 per cent with no ox. A recent survey gave the following percentages for the awraja: 7, 29, 37, and 27 respectively (MOA, October 1984:206).

The situation in 1986 was of course much worse as many peasants had lost livestock, including oxen in the famine and had not yet restocked. A great majority of the peasants I interviewed orally reported that they had sold one or all of their oxen in the preceding year.

There are a large variety of arrangements for renting (sometimes called borrowing) draught animals in Ambassel, of which the following are the most common. The list does not include group or communal cooperation (*jiggae*) discussed above, or outright charity in which a peasant loans his work animals to his neighbour for free, which was not uncommon in the first few months of post-famine recovery.

---

[1] Yeraswork Admassie, Sociology Department, Addis Ababa University, personal communication; also a computer print-out of enumeration in his possession. The peasant who told McCann that he had married ten wives in ten years was most probably pulling the ferenji's leg.

1. *Megenajut* or *mewaguir:* This is a practice in which two peasants with one ox each decide to team their animals and work their land in turns. No fee or remuneration, in cash or in kind, is involved here.

2. *Wonfel:* This practice involves renting oxen in exchange for labour. The arrangement often calls for two days of ploughing work on the land of the animal owner for each day the renter uses the oxen for himself. *Wonfel* is the most common form of renting farm animals in Ambassel; it is also quite widespread in the province, although it may vary in some minor details from place to place.

3. *Chinnet:* In this arrangement the peasant renting oxen agrees to pay in kind for the use of the animals at harvest time. What the payment will be will depend on how much land the user has ploughed and what he has grown. Two *dawulla* of grain for the use of a pair of oxen is considered a high price to pay (1 *dawulla* equals 1 quintal).

4. *Ye-Agbi* (*masgebat* means to train an ox for farming): A peasant may ask to train someone's young ox by working him in the field. The trainer will not have to pay anything in the first season of his training, but will be expected to pay in food grain if he uses the ox a second time. Usually the trainer will team up the young ox with an older, experienced one.

5. *Ye-Aggeda* or *Ye-Geleba* (*Aggeda* means stalk, *Geleba*, straw): Here the peasant agrees to offer the owner from whom he has rented the animals all the crop residue on his land after harvest.

For those unable to physically farm their land—the reasons may be due to old age, poor health or being a woman—arrangements can be made to have their plots worked by others in exchange either for a fixed payment, or a percentage of the harvest. Here too, the arrangements can be very numerous, depending among other things, on whether the labouring peasant brings his own oxen (or seeds) or uses those of his host, whether or not the agreement involves only ploughing, with other work to be carried out by the owner of the plot himself, etc. etc. In a crop sharing arrangement, equal sharing of the harvest between "labourer" and plot owner is considered the highest form of payment.

There are of course serious adverse consequences for the peasant with insufficient draught animal power; these consequences are further aggravated if such a peasant also happens to be physically unable to do any farm work.

To begin with, a peasant in this condition will not be in a position to work his land at the right moment and in the most efficient manner, and this will reduce his yield in the end. Secondly, a pair of oxen already fully employed by their owner—in a majority of cases, the owner always offers his animals for rent after he has completed his own ploughing—will have lost some of their draught capacity by the time the poor peasant gets his chance to use them. One writer has estimated that a pair of oxen in the highlands normally work some 450 pair-hours a year which is equivalent to 50 to 70 work days (Anderson 1985: 28). Sharing of oxen with someone means doubling the amount of time worked by the animals, and thus reducing their capacity by as much as a half. Thirdly, the poor peasant will not be able to plant what he believes to be the right crops or crop mix in a given season. Since he will work his land late, and since the reduced capacity of the draught animals he has rented will force him to select those crops which require fewer repeated ploughings or less soil preparation (ex. pulses), his harvest and his income will suffer at the end. Dependency here means not being able to work one's land skillfully and efficiently, and not being able to choose what to plant and when.[1]

Farming practices

The main crops grown in Ambassel are maize, teff, barley and horse-beans. Official crop surveys show that since the early 1970s sorghum (which is of two kinds, white—*mashilla*, and red—*zengada*) has gained ground over all other crops and now accounts for more than half of the total food grain produced in the awraja as well as the province. In the 1960s, the main crops were, in order of importance, teff, barley, sorghum and maize (Wollo Province Development Office 1968: 12ff). The ascendancy of sorghum since then has been mainly due to the fact that the crop is drought tolerant. Generally speaking long-stem or stalk crops such as sorghum and maize require less labour, and involve less cost than small-seed crops such as teff, barley and wheat. Crops in the former group can be stored or sold without threshing, winnowing or field stacking, and they are also known to sustain less post-harvest loss than crops in the latter group. The increased importance of sorghum

[1] In light of this it is encouraging to hear that some development agencies are engaged in experiments to ease the burden of the one-ox or ox-less peasant. For ILCA's single-ox plough experiments, see Gryseels *et al.* (1984), Gryseels and Jutzi (1986), Abiye Astatke *et al.* (1986).

and maize may be taken as evidence that the food habits of the population in Ambassel, as well as in eastern Wollo, is gradually changing.

We found in our field survey that a majority of peasants had planted at least four main crops in the current season, with a small minority (poor peasants mainly) a minimum of two main crops. There were no peasants who had all their land given over to one crop. A peasant divides his land into several plots, on each of which he plants different crops—a practice which may be called multiple cropping. As a general rule the peasant is guided in his choice of crops by the following maxim: first the necessary, then the obligatory, and last the useful. The needs of his family are primary, and the bulk of his harvest is consumed at home. But he will have to meet certain obligations to others, of which those to the state are by far the most onerous and unavoidable; but he may also have obligations to other peasants from whom he may have rented oxen, borrowed seeds or food grain. If there is a surplus after this, he will convert that into cash to buy livestock or other assets to improve his production or enlarge his savings. In both the last two cases, the peasant will be influenced to some extent by the behaviour of the market.

It may be worth noting here that Ambassel peasants do not favour mixed or inter-cropping, as opposed to multiple cropping. Mixed cropping is a practice in which a single plot is planted with two or more crops during a season. This technique, fairly common in west Africa, is widely regarded as advantageous to the peasant, particularly to the smallholder, as it is said to maximize efficiency in the use of land, provide opportunities for significantly increasing the variety and quantity of food produced, and reduce losses due to pests (Norman 1977, Taylor). Asked why the technique was not widely used in Ambassel, many peasants responded that shorter-stemmed crops in a mixed plot will be shaded from the sun by taller crops and will not be able to grow healthily. The only exception to the general disfavour of inter-cropping was the home plot.

The home plot, generally a small area, is often planted with stalk crops (maize was the main one in the autumn season in 1986), in between which one finds vegetables, pepper and spices. This is the only plot which benefits from natural fertilization since livestock are kept in the homestead and some of the manure is applied to the crops in the compound. Crops in the home plot are also better watered as water from washings and similar activities is thrown out in the compound. Ambassel peasants use neither chemical nor natural fertilization (manure) to enrich the soil of their main plots. Cattle dung, collected and formed into patties, is used almost exclusively as fuel for home use. The home plot, worked mainly by the women of the house, is used to grow seed for the next season, and secondary crops that earn the family, specially the women small spending money.

The chief form of reviving the fertility of the land is crop rotation, fallowing in most places of the awraja having been abandoned a long time ago due to the growing pressure on the land, and the increasing dimunition of holdings. Chemical fertilizers are neither cost-effective nor affordable by a majority of peasants.

There is nothing novel about the rotation system employed by farmers here; they follow the traditional and well-tried practices common all over the northern highlands. A peasant, mainly dependent on *meher* season (autumn) cultivation, will grow a legume on the average every two years on a given plot. The following is a common rotational formula employed by many peasants: legume—teff (or small-seed crop)—sorghum (or stalk crop)—barley—legume—quite frequently peasants plant a small-seed crop before and after a legume, but there are numerous exceptions to this rule. What may be new here is that peasants occasionally use soil crops, particularly *noug* (niger seed, *guizotia abyssinica*) as a soil enrichment crop in place of a legume. Ironically, the only time plots are "fallowed" is during drought when the land is unused, and when in addition the ground is covered with weeds and other vegetation (i.e. if the rains have not completely failed), and thus protected from erosion.

The agricultural calendar varies from area to area and from zone to zone, and a precise determination is not possible for an awraja as diverse and extensive as Ambassel. The *belg* (spring) season where in a normal year the rains are expected to last from mid-February to the end of April, comes to an end with the gathering of the harvest at the end of May or early June. This is immediately followed by preparations for the main season. Table 11 is a working calendar for the main season and the main field crops in the *woyna-degga* zone of our awraja.

Initial ploughing of course takes place much earlier than shown in the table. Experienced farmers plough their land soon after harvest when the soil is softer, and draught work is easier on work oxen. This is believed to contribute to a good harvest. Some peasants also practice what may be called dry-season planting, i.e. sowing seeds before the rains begin on dry-ploughed land. This, accordingly to some of our

Table 11. *Agricultural calendar for main crops in Ambassel*

Crops	Last ploughing before sowing	Sowing	Harvesting
Sorghum	March–April	May	Nov./Dec.
Maize	March–April	May	September
Teff	Early June	July/Aug.	November
Barley	July	July	October
Horsebeans	July	July	November

Source: Oral interviews of Woreda Extension Agents.

informants, will result in earlier maturation of crops and a higher yield. But both practices involve a strong element of risk: the first practice will expose the soil to erosion by removing the stubble cover, and the second may result in loss of seeds if the rains fail. Many farmers habitually sow soon after the first few days of rain when their plots are covered with newly-sprouting weeds which they eliminate by a final ploughing.

There are few crops which do not require multiple ploughing before sowing is finally complete. Teff (more than 5 ploughings), and following that barley and wheat require more ploughing than stalk crops. According to extension agents in the awraja, a pair of oxen in relatively good condition will take three to four days to plough one hectare of land in the *woyna-degga* zone, and more in the *qolla* zone. Weeding, which is the other arduous task in peasant agriculture, becomes less and less of a problem as one moves to higher elevations. Our estimation is that about 35 per cent of a peasant household's time in the *woyna-degga* zone of Ambassel is devoted to land preparation and seeding, another 20 per cent each to weeding and harvesting, and the rest distributed among the less strenuous tasks such as threshing, winnowing, transport and storage.

In the past, i.e. before the land reform, it was quite common for Ambassel peasants, especially those in the *degga* areas, to work one or more additional plots situated in different agro-climatic zones. This was to minimize the risk of crop loss due to natural causes, and to complement their family's food resources. The main natural hazards that give rise to widespread crop losses, then as well as now, are frost (in the *degga* zone mainly), water-logging (in the *woyna-degga* zone), and pests, specially locusts (in the *qolla* and *woyna-degga* zones). Land reform has now dealt a serious blow to this complementary farming practice although it has not altogether abolished it. At present the chances for complementarity exist where a peasant, who is unable to farm for physical and other reasons, rents his land to someone from another zone in exchange for a share of the harvest.

Degga peasants, specially if they happen not to be blessed with sufficient land and/or sufficient traction power, hire themselves out to (mainly) *woyna-degga* peasants, i.e. those with large enough holdings, as labourers during weeding and harvesting; if payment is involved it is almost always given in kind. The practice of hiring labour for a cash wage is virtually unknown. Often enough in fact a good number of peasants who work as temporary hands are related by blood, marriage or religious convention to the owner of the farm, and may be working for him not for immediate remuneration but for a favour to be received at a future date.

The division of labour follows the traditional pattern: it is structured along lines of gender and age. Everybody who is capable of working is expected to work, and labour is allocated to each family member in accordance with his/her age, experience and condition of health. Paradoxical as it may sound, the old 19th century utopian-socialist maxim—he who shall not work shall not eat—is more seriously adhered to here in rural society than in other sectors of the population. Women participate in all aspects of farm work except ploughing, and sowing of field plots. They are also active in marketing, more so than men, although they specialize in buying and selling low-priced products. A heifer, bull or ox is sold or bought by the man of the house and not by his wife who, in contrast, is free to deal in young sheep and goats.

There is an interesting aspect of the sexual division of labour which has an important bearing on the subject of survival strategies. The most important area in which women have exclusive sway is food processing (mentioned above), and *family food management.* By this we mean the allocation of food for each meal and for each day and the distribution of the food among all members of the family. The distribution of food is done on an equitable basis (equity as defined by customary values), and taking into account the age and physical output of labour of each person in the household; sick people and people of advanced years stand as exceptions to this rule. Now, this critical task of women—and we believe the same is true of women in many other rural societies—ensures that they cannot and will not be victimized during food shortages by their husbands or menfolk, as has been argued in the famine and related literature (in addition to the sources noted earlier, see Torry

1986, Kynch and Sen 1983. We shall return to this issue later). Indeed, during the early and middle periods of famine a woman's food management skills take on added importance, and the menfolk gradually become dependent on her ability to make the available stock of food last as long as possible.

It is interesting to note that recent developments in literacy and education, and post-revolution activities having to do with rural military mobilization have given rise to tendencies which in the long run will have consequences on rural labour and production. As a result of the expansion of educational activities, more and more young peasants are spending more time in school and away from farm work. Once these youngsters reach a certain level of school education—middle school and above—they develop an aversion, sometimes bordering on contempt, to agricultural labour. This is more often the case with boys than with girls.

The same thing is true of peasants who have been involved in military activities in one form or another. A peasant who had participated in the Ogaden Campaign in 1977–78 (as part of the peasant levy of that year), and whom we interviewed near Woldiya (in Yejju awraja) told us that he wants to become a rural mechanic, and was not interested in going back to farming. He was demobilized in 1980 but has not shown any inclination to join a PA since his return, and survived by doing odd jobs around the offices of the Woreda MOA and the Woreda PA. As these developments take out more of the energetic elements from the rural labour force, peasant families may take more stringent measures to retain their youngsters in the rural areas. Ironically, peasants often wish their children to take on urban-based, non-agricultural employment which they believe is more secure and more respectable.

The control of the micro-environment

It was noted earlier that peasant agricultural practices are in the long-run environmentally harmful. This must not be taken to mean, however, that the peasant has not evolved techniques to try to deal with the adverse consequences of his own acts. Contrary to some commonly-held views, the peasant is aware that the vitality of the land is slowly but steadily being drained away by the struggle for subsistence, and that he must not only do something to stop or reduce this loss but also attempt to rejuvenate the soil. The preventive and revitalizing techniques he has at his disposal have been tried and tested through many generations, and though by no means perfect or scientifically sophisticated, are fairly effective in the short-run, and the best that can be expected from a mode of production functioning within narrow bounds. Every farm reflects a considerable degree of awareness and a high level of intelligence and if only his detractors were conscious of this fact, the peasant would long have received the credit he deserves. (For the high level of peasant skills in West Africa, see Adams, 1986.)

Most farms in Ambassel are skillfully laid out and carefully and intensely worked. It is worth noting that the level of skill and agronomic knowledge, as reflected in the lay out and content of a peasant farm, improves (and sometimes dramatically) as one moves from the lower to the higher elevations. Indeed, some peasants, mainly those in the *degga* areas, were proud of their competence and knowledge, and held attitudes ranging from contempt to pity towards peasants in the *qolla* zones where farms are poorly kept, and the standard of crop husbandry is relatively low.

Peasants in the awraja are aware that removing the vegetation cover from the land exposes the top soil to erosion, and leads to the loss of stored moisture through evaporation or run-off. Traditional peasant practices, which are geared to the conservation of water and soil, attempt to bloc run-off and loss of moisture, and in the process prevent soil from being washed away. Some studies have made a distinction between soil and water conservation, and have argued that the peasantry in Wollo and other areas consider soil conservation as the only worthwhile conservation activity (Yeraswork and Solomon 1985: 26ff). In many respects the distinction between one and the other is a very thin one, as generally the loss of soil leads simultaneously to, or is accompanied by the loss of moisture. It is of course true that water conservation involves also the construction of large water retention systems, such as ponds, canals, artificial streams, etc., in which case the problem becomes different from that of soil conservation.[1]

We estimate that more than half the total annual run-off in Ambassel (and this may be true for Wollo as a whole) occurs in the month of July when perhaps not more than 30 per cent of the season's rains have

[1] It is interesting that nomadic pastoralists like the Borena and Afar employ more refined water conservation techniques than highland agriculturalists (see Sandford 1983: Ch. 2).

fallen. This is the period when the land, after having been repeatedly ploughed, has been seeded but remains naked and unprotected. Peasants know that it is this post-ploughing condition of the land which exposes it to the greatest danger of erosion. The softness of the soil itself at this time considerably reduces the ability of the earth to retain water. Peasants here employ several indigenous techniques to control the micro-environment and to maintain the fertility of the land. We have already discussed peasant practices regarding the rejuvenation of the soil; these practices have now narrowed down to crop rotation, as manuring and fallowing have been given up for the reasons noted above. The conservation techniques discussed below are employed, specially by peasants engaged in slope cultivation, singly or, as is often the case, in combination.

Most peasants plough along contour lines, and this is the simplest form of blocking run-off; ploughing in the opposite direction is usually done to drain the land of water if the threat of water-logging is serious. Peasants also build trenches around their plots for drainage purposes. *Contour terracing* is also a traditional art long practiced by the peasantry, although at present hundreds of kilometres of terraces have been built by peasants mobilized by their PAs at the initiative of the Ministry of Agriculture; in many cases payment is made to participating peasants through the Food-for-Work Programme of the UN's World Food Programme (WFP). Almost all of the modern terracing (called step or bench terracing) is made of stones which some peasants disapprove of because they say it easily harbours rodents and makes it difficult to get rid of them.

Another traditional art which is now almost lost is *irrigation.* At present Ambassel peasants draw water through tiny canals on their land only at flood times, i.e. when a water source—a river or stream—overflows its banks. Drawing water at non-flood times through a system of large canals and trenches, and controlling its flow and regularity is now a lost art in the awraja; so too stream diversion or artificial ponding. Cossins found evidence of a fairly complex and large-scale irrigation system of considerable antiquity in western Wollo in 1973 (1974: Zone 2, 48–50).

Large *bunds*, built out of stone or earth, are also commonly employed by the peasants to control erosion and run-off. One study found in fact that sampled peasants in Yejju considered bunding the most effective form of controlling erosion (Yeraswork and Solomon: 30ff). As in the case of terracing, some peasants preferred soil bunding to stone, for the reasons noted earlier, and also because the former is easily and quickly covered with vegetation.

Yet another technique quite popular in Ambassel is *contour ridging*, often combined with *furrows* or ditches dug again along the contour. This technique can allow the storage of considerable amounts of moisture which would otherwise be lost by run-off and evaporation. A variation of this technique is to allow grass or at times hedges to grow on the spine of the ridges built in between plots. The furrows are used to hand-plant seeds during a short *belg* or *meher* season. Since the furrows retain moisture long after the rains have gone, crops planted in this way have a better chance of survival. Ridges covered with hedges have the added advantage of breaking the force of the wind which in certain months in the temperate and colder areas is strong enough to damage crops.

Hedges in between plots are also not uncommon, and here again the purpose is to control run-off and erosion. Since land is scarce the hedges have to be thinly planted following the contours of the plot. The taller the hedges are, the better. As a general rule a peasant combines ridging with hedging, that is, if he has built ridges he prefers to grow hedges on top of them rather than separate the two techniques on the same plot. This saves land for cultivation. Related to this is the technique known in the literature as *strip cropping*, a practice where a thin strip of grass is grown in between plots and the land ploughed and cropped without touching the grass strips. There were a few peasants in central and southern Ambassel who had planted rows of trees in place of grass between their plots, but most probably this was the influence of extension agents rather than a product of indigenous agronomy.

What about the considerable amount of soil and moisture loss that occurs during seed-bed preparation, and in the actual process of ploughing? The two most common techniques employed by the Ambassel peasants to control loss during this phase is *ratooning* and *mulching*. In the first case plants are cropped above ground level during harvesting, and are allowed to sprout again. The roots of such plants continue to hold the soil and the new plant will mature without too much soil preparation work. Ambassel peasants actually practice partial ratooning: crops, often stalk crops like maize, are cut just above ground level at harvest time and the land ploughed without uprooting the stalk as long as possible. By the final ploughing, however, the last remains of the older crop will have been removed, although the stalk and root residues are left lying on the ground. The purpose is to prolong as much as possible the moment when the land is completely unprotected.

Mulching involves the use of crop residue and other vegetation to cover the land during ploughing. This is to protect the topsoil and to reduce moisture loss. Vegetation mulch will not remain on the land for long as it is used to feed livestock; some peasants prefer stone mulch instead, but to get hold of a sufficient number of stones one's plot will have to be located near a stony area.

A novelty specific probably to Ambassel and eastern Wollo peasants is the *use of noug* (niger seed) not just as a soil enrichment crop (akin to a legume) but as a soil conservation device. This crop, which is indigenous to Ethiopia and is a favoured oil crop in the area, has a number of properties which the peasantry exploit fairly well. First, its roots are believed to hold the soil well, and for this reason peasants working on slope land which has a step (or bench) terrace, plant noug on the upper and lower edges of the plot to reduce erosion. Secondly, noug is repellent to rodents, and is not favoured by crop-damaging livestock such as goats, and hence peasants often plant the crop on all sides of a plot to serve as a fence around another crop. Planted in this way, or (if the peasant does not have sufficient noug seed) planted on the windward side of a plot, it acts as a windbreak for another plant, usually seed plants like barley, wheat or teff. Noug is planted usually in June and harvested in mid-to-late December, and peasants often grow it together with barley, horsebeans, wheat, and at times teff to allow the land to be partially protected when the main crops are harvested in October and November (see Table 11 above for crop calendar). This is the closest Ambassel peasants come to what in the literature is called *serial cropping*, a technique in which an additional crop is planted into a maturing one to prevent the land from being cleared of all vegetation.

Noug is often ratooned in the manner discussed above. Since it requires very little land preparation and virtually no weeding, the crop is highly favoured by peasants. Its bright yellow flowers attract bees and farmers in the areas where the crop is most cultivated practice apiculture. As an oil crop, noug of course has high cash value.

Green technology techniques are not employed by the Wollo peasantry, nor are such techniques encouraged by extension agents. The only new ideas pursued by MOA agents are chemical fertilization and the use of improved seeds, and the only new conservation activities such agents are involved in are reforestation, the building of check-dams, and river diversion and spring or water development work (wells, etc.). Clearly then, with his limited capabilities the peasant is far more resourceful and innovative than the state agencies in the field with their vastly superior income, expertise, and extensive international support in the area of environmental control and soil conservation.

Peasants have a good knowledge of seed varieties and keep as many different types in store as they can; this enables them to quickly change their seeding plans in response to changes in the weather. The *belg* rains are notoriously unreliable in most parts of Wollo, and should there not be sufficient rains in a season peasants will want to try short maturing varieties in place of the normal ones. The sorghum planted in the *qolla* zone (called *degalitu* in Ambassel woreda, and *gedalit* in Tehulederae) is different from that grown in the *woyna-degga* zone. The normal variety matures in 6 to 7 months, but there is a short-maturing one, called *jammiyae* which takes about 3 months to mature. *Jammiyae* is very popular with Ambassel peasants; it can be sown as late as June or July and harvested in December, and serves as insurance against the adverse effects of pitiless nature.

Similarly peasants keep at hand fast-maturing varieties of maize (called *limmat*, or *fatina irrie*, this meaning strong-muscled arm in Oromigna), and also of teff (called *birgae*), both of which have a maturation time of approximately three months. Peasant informants explained that some legumes, particularly horsebeans, are useful plants in stress conditions because they have relatively low moisture requirements, and the early maturing strains can be sown and harvested in less than four months. Interestingly enough famine reserve crops, that is, those that grow in difficult conditions and require very little moisture, such as potatoes and root crops, are not widely grown in the awraja. A few peasants in Worebabo woreda raise taro which they eat only in times of serious food shortages.

Rodents and pests are a major hazard to crops in both Ambassel and Wollo. At the time of our visit there was great concern about the dangers of pests and rodents among peasants and extension agents. The main rodents are field-mice and one peasant informed this writer, half joking it seemed, that he was considering raising cats to try to drive away the mice from his fields. A far more serious danger was however the quelea bird which feeds mainly on sorghum. Peasants spent hours standing on raised platforms with a sling and stones in their hands guarding their sorghum fields against these birds. These peasants were well-aware of the feeding habits of the bird which attacked crops mostly during the morning hours and at sundown.

The most common pests at the time of our survey were American bollworm, armyworm, various kinds of stalk borer (a danger to maize and sorghum), aphids, and grasshoppers (for a discussion of recognised pests and rodents by region see EPID 1974: 48–51; also Crowe and Shitaye 1977). There is no reliable measurement of the extent of crop loss suffered by peasants due to pests and rodents. One report (EPID 1974: 36) suggested 20 per cent of the harvest is lost in this way in an average year nationally. Pest infestation is very high in the first years of famine-recovery and declines gradually after that. This explains the apprehension of peasants in Ambassel in the autumn of 1986 which was their first normal season in about two years. It is our estimation that in an average year peasants here may lose 10 to 15 per cent of their crops to pests and rodents. No peasant can afford the cost of pesticides, and very few peasants in the area were known to have used them.

There are basically two kinds of grain stores employed by peasants here. For small-seed crops (teff, barley, wheat, etc.) and pulses, the *gotera* is generally preferred. This is made of wooden frames, leaves and twigs, and plastered with mud. It is placed on a raised platform and kept inside the compound of the homestead. A smaller *gotera* for similar crops may be kept inside the house. The second type is the *goudguad* discussed above. This is mainly for strong sorghum and maize, that is stalk crops which may be stored without threshing. According to the several informants crops here are protected from pests and insects in the following manner: before the *goudguad* is ready for use the leaves of the *endod* plant (*Phytolacca Dodecandra*) are scattered in it; a day or so later, ground red pepper is sprinkled on the inside walls. The treatment is repeated a second time if the owner is not satisfied. Crop loss in storage is also another subject on which we have very little information. Peasants are reluctant to divulge anything about the content of their stores, the value and effectiveness of which is thus difficult to assess.

Consumption, marketing and prices

A great number of peasants here practice bimodal cultivation, however, the unreliability of the *belg* rains is such that for most practical purposes it is mainly the autumn season which provides the average peasant his principal income. For almost all peasants their source of income is agriculture, livestock and livestock products. Asked about side-line occupations, about 94 per cent of our respondents said they had none; only 6 per cent reported that they engaged on a regular basis in non-farm income-earning activities which were mostly handicraft work and petty peddling. Most peasants were proud to be nothing but farmers, and handicraft production or other side-line activities, except perhaps trading, are looked down upon which makes peasants who practice toolmaking, weaving, etc., reluctant to admit them. "*Mekessem bicha*" (doing only farm work in Wollo Amharic) was the proud answer we frequently got to our questions.

The craft industry of Wollo and Ambassel is relatively underdeveloped. The most frequently practiced rural crafts in the province as a whole were weaving and smithing. In addition, in Wore-Illu and Wadla Delanta awrajas carpet making has become an increasingly important source of income for peasants. In Aussa and Qallu awrajas, jewelry making, particularly gold and silver smithing, is a side-line occupation with a long history, although most of the skilled jewelers are located in urban centres. We should note also that both awrajas have large pastoral populations, and that this craft is not found among the sedentary population in Wollo if HASIDA's records in Dessie (October 1986), on which the discussion in this paragraph is based, are reliable. In Wore-Himenu, women peasants earn some side-line income by making embroidered cloth (*tilf sirra*).

Before we move to the main subject at hand one other problem should also be dealt with, namely post-harvest loss. There are a number of factors that have a bearing on the magnitude of post-harvest loss among peasants but these differ from place to place. In Ambassel, the main ones were: the skill of the farmer, especially in work involving cutting, threshing, etc., and how quickly crops are removed from the farm and taken to storage; the physical location of the plot, i.e. whether or not it is easy for rodents and livestock to damage the crops; the distance of the plot from the homestead, and the manner in which the crop is transported. Some official reports indicate that on the average a peasant sustains a 10 to 20 per cent post-harvest loss every season. This we believe is quite exaggerated, and does not take into account the various factors in each region that help to reduce loss. Some recent studies are beginning to cast doubt on the generally accepted view that post-harvest loss in many rural areas is typically high (IDS Bulletin 1982). We were unable to measure farm-level losses as most of the fields were not harvested yet, nor were we able to get access to reliable information on the subject. Nevertheless, we estimate that peasants in *woyna-degga* and *degga* areas of Ambassel sustain on the average post-harvest losses of less than 10 per cent, and perhaps as little as 5 to 7 per cent.

With this and the earlier discussion in mind, let us examine the level of material welfare of the peasantry, taking, for analytical purposes, a household of five, with a land area of 1.5 ha of which 1 ha is devoted to cereals and the rest to pulses; such a household is not untypical in Ambassel. We shall use crop yield and marketed surplus figures given in the 1979/80 agricultural survey (MOA-CSO: 77, 125), because the harvest of that year was good in Wollo and it was the only time estimations for marketed surplus was made by either MOA or CSO surveys (except the 1983 MOA survey which however is defective because it was carried out at the time of the drought); moreover rural data collected after 1980 is distorted by the dispersing food crises in the area.

According to this survey, the average Wollo peasant had a crop yield of 14.7 quintals per hectare (qn/ha) of cereals and 12 qn/ha pulses in that year. A peasant with 1 ha and 0.5 ha of cereal and pulse land would thus have 14.7 qn and 6.0 qn of cereals and pulses at the end of the main season. Let us disregard post-harvest loss for the time being and assume loss to pests and rodents to be 10 per cent. According to the same survey a peasant marketed 12.42 per cent of his cereal harvest and 11.4 per cent of his pulses to buy necessities such as salt, oil, paraffin for lighting, etc. We shall assume a very modest seed requirement rate of 10 per cent for both crops. Finally the peasant payed in 1986 a tax of 20 Birr, a famine levy (see below) of 20 Birr, and obligations to AMC amounting to 0.25 qn; the first two add up to 1 qn of cereals and 0.25 of pulses. At the end of the main season his total expenses will come to 5.8 qn of cereals and 2.15 of pulses, and he is left with 8.9 qn and 3.85 qn of the two crops for home consumption.

Another way of putting it is that the peasant will spend 40 per cent of his harvest in cereals, 36 per cent in pulses to meet his recurrent obligations (not counting membership fees to rural mass organizations, and the like). It is generally accepted that to acquire the necessary calories and energy a peasant family will need to consume at least 12 qn of cereals and 6 qn of pulses a year. Under the circumstances peasants have two options: they may either try to cover the deficit by selling livestock, mainly goats, or forego the temptation and remain with reduced consumption capabilities until the *belg* season, hoping to make up the shortfall then. The reader should note that we have used high crop yield figures obtained in one of the best seasons for Wollo in recent years. Less optimal weather and farming conditions will reduce production and yields sharply.

The formula we have used is admittedly unsophisticated but it is sufficient to show that Ambassel peasants as well as peasants in the northeast region as a whole are, even in normal conditions, and in times free of environmental and social stress, not too far removed from the precipice of starvation and death. Since, as noted earlier, the area's developmental prospects are poor, and will become poorer in the years to come, and as the system of peasant production practiced here has been in serious decline for decades, the fate of the peasantry of the region is more frequent and more intensified food crises. It may be said of Wollo in particular that famine is hiding behind the rugged mountains of the province and will strike again before the people have fully recovered from the disaster of the past.

Markets and marketing

We shall not attempt here a full-scale discussion of markets and marketing as this will take us too far from our subject. We shall instead deal briefly with the main elements of the rural market that have a bearing on famine response. We shall also not treat livestock marketing here as it will be dealt with in some detail in a later chapter.

Markets play an important role in peasant economy and social life. They are the means by which a peasant household manages its resources, and resource management here means selling one's surplus to make up one's deficit, and balancing the family's basic needs. The rural trade system is an intricate one, and the average peasant will have several markets to choose from, and may periodically travel between 10 and 15 km to attend the market of his choice. In exceptional circumstances, particularly when the markets in his area are behaving unfavourably, a peasant may take a long journey—perhaps lasting two or more days—to try his luck in a large market located in a different region. Several of our informants in Ambassel told us, for example, that they travelled to Wore-Illu awraja in western Wollo (a distance which would perhaps take a peasant travelling on foot several days) to sell livestock and to buy grain during the famine. They said prices of both livestock and food products were much better there than in their own localities at the time. In normal situations, however, markets are basically a local affair.

Peasants also attend markets to meet relatives and friends, to exchange information (about agricultural conditions, market behaviour, seasonal employment opportunities, etc.), to discuss exchange arrangements (livestock loans, land renting, and the like), to make social deals (marriage, etc.), and to have a little bit of entertainment. In Haiq and a few other major towns in our awraja, "entertainment" included, especially among the younger peasants, having group fights with sticks, these fights got to be such a danger to people and property that urban defence guards now disarm all peasants of their weapons on market days.

Peasants do not have access to the media, and information on social, economic and political matters in other areas is disseminated in the rural areas through the market system. This information is just as important to the peasantry as it is to the government bureaucrat. A peasant household's resource management decisions may be seriously affected if for example a disease (known or unknown) is spreading in one neighbouring community, if there has been civil conflict in another, or if crop conditions in a third are expected to be bad.

Markets vary in serveral ways: in size, in the kind of buyers and sellers they attract, and in the range of products that are marketed. By markets we mean those that are held weekly, and which attract at least several hundred people; small informal neighbourhood markets, often held more than once a week in a *mender* or *got*, are not included.

We may distinguish three types of markets and three levels of exchange. The first may be called the *local* market. Most markets held in woreda capitals fall into this category, but there are also small towns and villages where such markets also take place. In the Ambassel area markets also take place. In the Ambassel area markets of the latter type are held at Rob Gebeya (a short distance north of Dessie; it is not in the awraja but attracts a good number of Ambassel peasants), Golbo, Adami and Gerba. The chief actors here are peasants, and exchange takes place for the most part between one peasant and another. But there are always a small number of rural traders, both those who come to sell urban-made goods (cloth, footwear, etc.), and those who buy peasant products. In a good number of cases rural traders may be peasants themselves who engage in such work on a seasonal basis.

The second kind of market is the area market which is larger and more diverse than the local market, and attracts rural traders and merchants. Peasants sell mostly to the latter two, and professional merchants buy from both peasants and petty traders. Prices are higher by a small percentage here than in the local markets. Often enough the awraja capital serves as the main area market but this is not always the case. For southeastern Wollo, for example, Woldiya, Haiq and Bati serve as the most important area markets.

Dessie and Kombolcha may be taken as Wollo's *regional* markets, the third kind of market. Here the influence of the merchant is more pronounced and may be compared to that of the peasant on the level of the local market. Often enough the peasant, who may intervene at any level in the market-system, comes to the regional market to buy products that require large outlays, such as farm implements and their accessories, manufactured goods, etc. Regional markets offer greater choice than the other two. Both Dessie and Kombolcha are however not well-developed compared to regional markets in other provinces.

There are two other important factors that make the rural market system far more complicated than it appears. First, in most cases each market in each category has some small attraction to offer the visitor; for this and the other reason noted below, many peasants do not depend on one market but attend instead two or more markets regularly. As a result of its physical location, the favoured agricultural product of the people nearby, and the side-line and non-agricultural specialities of the area, a market may be noted for a certain individual product or class of products. Wichalae is a good local market for certain cereals and pulses similarly Golbo is noted as a market for livestock, especially mules and horses, while Haiq is an important cattle market. On the other hand, many of the markets we visited in the awraja were poor in traditional implements and leather-work for which peasants probably travel to Dessie, or north to Woldiya.

The second factor has to do with what may be called *food complementarity*. It is often the case that food and other agricultural products are traded from surplus to deficit areas. There are, as noted above, a number of areas in Wollo which are often surplus areas and food moves from them to the deficit areas. However, there are within smaller regions like Ambassel areas which may be food surplus in one year and deficit in another; here the direction of food exchange is not always the same, and will vary from season to season. Furthermore, an area may have a surplus in one or two kinds of crops and be in deficit in other crops. In this instance there will be a two-way food flow as peasants will exchange what they have for what they do not have with peasants from other areas (for a discussion of food complementarity in an African setting see Pottier 1986).

But food complementarity is made more complex by the fact that a peasant grows only a limited number and variety of food crops in any given season. The limitation is imposed by the smallness of his holdings, and by the specific ecology in which he lives. Note the variety of crops grown at each of the three main ecological zones in Ambassel shown in Table 3 earlier. It is the market which allows peasants who need one or more food crops, or who are placed in different ecologies to dispose of their surpluses and to make up their deficits. A *degga* peasant may travel to a market with his barley which he sells and with the proceeds purchases sorghum or noug, both of which he needs but does not himself grow. The exchange is carried out at local and smaller markets often among peasants themselves; this involves of course travelling to different markets in different areas. At times however it may be mediated through the rural trader who brings the products of one zone to the markets of another.

Now, during the onset of food crisis a number of developments take place which have a bearing on the behaviour of markets. First, peasants reduce the range of food crops they normally consume to a few essentials, and hold other crops they may have to sell in the market. Thus, in the early phase, those crops that are not regularly eaten as food, like barley, wheat, oil crops, etc., are traded, and staple food crops (sorghum, teff, pulses, etc.) are eagerly bought; some peasants may dispose of one crop, say maize, if they have another crop similar in food value, e.g. sorghum. What a peasant chooses to sell and to retain will vary of course from household to household. This increased marketing activity may not be reflected in any significant price movements up or down as one will cancel out the other. Secondly, peasants will travel to a large number of local (or even smaller) markets in search of specific food crops, of favourable bargains and of information about markets and price behaviour. Third, the rural grain trader assumes an added importance now as he brings scarce commodities from far off areas. The large markets, that is the area and regional ones, will begin to show that a crisis is looming only after the local markets have been exhausted of most kinds of food crops. Thus, market or price behaviour is not a good indicator of what is happening in a rural area, as some have tried to argue (Culter 1984), until what is actually happening is already evident.

A peasant often has a choice of several markets that are located within a reasonable distance of his homestead. In an area as large as Ambassel there are markets in one part of the awraja or another on almost every day of the week, except Sundays in Christian communities or Fridays in Moslem ones. Most markets do not as a general rule compete with one another, rather they complement each other. This is to say, except for distant markets, not two markets in one or neighbouring local communities are held on the same day.

A peasant living in the Haiq area will attend the market there on a Friday, and if he is not satisfied with the prices offered, he may take his products to the Golbo market which is held on Tuesdays, and if still unsatisfied he may try Rob Gebeya market which falls on Wednesdays. Similarly, the peasant living in Wichalae near the border with Yejju has the Wichalae market (Mondays) and the Girana (Thursday) market across the border to choose from; there are two markets just south and west of Wichalae which the peasant can also attend.

The greater the number of markets a peasant attends the more informed he will be about prices, the movement of goods and the range of products currently in demand. Most peasants, including peasant women, follow closely the movements of prices in their areas, and some in distant ones as well.

Since 1979, when the government, through its purchasing agency the Agricultural Marketing Corporation (AMC), set fixed grain prices, peasants are offered two sets of prices for their food products. Each peasant has to deliver through his PA a predetermined quota of grain to AMC every year. How much a PA's, and therefore an individual peasant household's, annual quota will be is decided by AMC bureaucrats, and quite often in an arbitrary manner. The amount a peasant delivers to the state purchasing agency changes from year to year but the grain price rate he is offered by AMC remains the same. What the agency collects from the peasants is taken out of the rural areas, and transported to the major urban centres, especially Addis Ababa, to feed the population there.

If there is one government agency that is highly unpopular in the rural areas it is AMC. Peasants are resentful of its exploitive practices, its authoritarian ways, and its insatiable appetite. AMC is basically engaged in forced procurement, the purpose of which is unclear to many peasants; every year it takes food away from those who produce it to give to those who do not. As one disgruntled young peasant near Woldiya put it, the agency seems to be engaged in robbing the poor to feed the rich. Interestingly enough, several reports commissioned by the agency itself (AMC July 1985; also World Bank 1983) are critical of its requisitioning practices and yet neither the policies nor method of work of AMC has changed for the better over the years.

Any surplus the peasant may have after meeting his obligations to AMC may be sold on the market where prices are generally significantly higher. Table 12 provides comparative prices for food grain in several markets in Wollo province.

Table 12. *AMC and free market prices of grain (Free market in Ambassel and Qallu awrajas) (In Birr per quintal or 100 kg)*

Crops	AMC price to peasants[a]	Free market prices[b]	
		June 1986	Sept. 1986
Sorghum (2 grades)	27, 31	71.80	54.61
Maize	24	63.50	43.78
Teff (3 grades)	39, 43, 50	119.63	100.73
Barley (3 grades)	30, 30, 32	70.65	46.19
Wheat (3 grades)	34, 35, 39	74.30	49.87
Horsebeans	30	84.45	84.30
Fieldpeas (2 grades)	36, 40	108.00	88.85

Note: Free market prices for high grade crops only.
Source: [a] AMC (June 1985:63).
[b] From the records of the Red Cross, Dessie, October 1986. Average prices from markets in Ambassel and Qallu.

The reader may note that free market prices for some crops are nearly three times as much as AMC prices. It should also be noted that prices in the rural markets fluctuate during the year while AMC prices are unchanging and have remained fixed since they were set in the early 1980s, clearly a disadvantage to enterprising peasants who otherwise would hold their crops until mid-year when prices begin to go up sharply.

AMC quotas were not lifted during nor soon after the 1984/85 famine in Wollo; they were reduced for the 1985/86 crop year from an average of about 25 kg (the average in Ambassel) to 10 to 15 kg per household. A great majority of peasants were unable to provide the grain imposed on them, indeed, as discussed earlier, a large number of them were still dependent on relief assistance. Despite the considerable protest from extension agents and some MOA specialists in the province, AMC authorities remained adamant that peasants should deliver their quotas on time and at the right collection centres. *A number of peasants told us that they had to sell livestock or other possessions to buy the food required of them on the free market and at free market prices, and to deliver it to the authorities at AMC prices, involving a loss of anywhere up to 300 per cent.*

To appreciate how great the loss incurred by the peasantry in the exchange system is, Table 12 should be considered together with the movement of prices of consumer items needed or purchased by the rural population; many of these items are supplied by government agencies. Since current information on the subject is difficult to get hold of we have used data prepared in the early 1980s.

Table 13. *Price rises of peasant consumer goods, 1978–82*

Peasant consumer items	Price index % increase 1978–82
Cooking Oil	140
Soap	208
Canvas Shoes	174
Khaki Cloth	140
Sugar	157
Fertilizers	177

Source: AMC July 1985:19.

Before we end this discussion, a few words about other obligations the peasantry was burdened with at the time of our field work. *All through the food crisis of the previous year and after, peasants were expected to pay their taxes as well as their fees to the rural organizations of which they are members.* Furthermore, and this came as a surprise to us, each peasant had to pay what may be called a *famine levy*. In October 1984, the government, after formally admitting the existence of famine in the country, ordered that each Ethiopian national should pay a set sum of money *to help victims of the famine*; salaried persons were asked to give up one month's salary, while peasants and others were made to contribute a rate of payment determined by local authorities. This was the famine levy of 1985/86.

In Ambassel and the adjoining awrajas the famine levy was set at *20 to 25* Birr per peasant household, and had to be paid (for administrative reasons, presumably) together with the tax. Some peasants who expressed an opinion on the subject said they failed to understand why the very victims of famine should be forced to pay a famine levy.

Mention must also be made of a wide range of "campaign" and "corvée" labour demanded of peasants. Communal projects which required large outlays of labour, such as large-scale terracing and the like, were done by peasants mobilized by their PAs. In some cases payment (in kind) was involved, in others the work was done free. The land of peasants on military, or other state-sponsored duty was cultivated by peasants who were assigned to work on them in turns by their organizations. In brief, *a peasant devoted at least one day in the week on communal work and work on other people's land. Additionally, at least once a year peasants in Ambassel and the surrounding awrajas had to spend several weeks working in the plantations in the lower Awash, in Aussa awraja. This was in effect forced labour for the peasant was not paid for his work but provided only with food during his stay.*

So much in brief about peasant production in Wollo and the systemic as well as externally-imposed constraints under which it is now functioning.

As was stated above, *the seeds of famine are already being sown now, at the moment when recovery is under way*. The indigenous conservation techniques discussed above, while effective in the short-run, are totally inadequate in preventing or reducing long-term resource degradation. As each peasant breaks the soil, plants his crops, and brings in the harvest at the end he helps set in motion a chain of natural and social activities which in the final instance will lead to the loss of fertility of the land and of the vitality of the peasant household. Each year the Ambassel as well as the Wollo peasantry will grow less, "save" less, consume less, and in general become poorer and more dependent on optimum farming conditions to eke out a bare existence. The crisis of livelihood in Wollo is a profound one, and it will in the future be enough for minor adversity to unleash the destructive forces of hunger and pestilence.

6. Famine in Wollo

> In the case of famine, ... as in that of war, the occasion of the breakdown must be distinguished from its causes. The former is commonly a failure of crops caused by drought or flood. The latter, though aggravated at present by political anarchy, consists in the primitive organization, and absence of surplus resources over daily needs, which turn the misfortune of individuals into a general catastrophe... Famine is, in short, the last stage of a disease which, though not always conspicuous, is always present (Tawney 1932: 77).

Tawney made these observations about China before the revolution, but he might just as well have been speaking about Wollo before and after the revolution in this country. Indeed as in China, there are parts of Wollo or northeast Ethiopia from which famine has rarely been absent in the last two hundred years if the meaning of the word is a shortage of food leading to starvation on a sufficiently wide scale. It may be stating the obvious but serious crop failures leading to famine have occurred in Wollo caused by drought, too much rain, locusts or other pests, frost or hailstorms, human or cattle disease, and war and social destabilization.

There is sufficient evidence to indicate that the frequency of famine in Wollo and the northeast has increased markedly since the beginning of this century. The earliest food crisis that some of our informants in Ambassel and Dessie Zuria awrajas could remember was that which occurred soon after the Italian war and the occupation, probably in the 1936/37 crop year. One is not fully certain but the food shortages that preceeded this may have taken place in the latter part of the first decade of this century, perhaps between 1915 and 1918 (C.A. Wood, Pankhurst 1968: 220). McCann (1984: 158ff) has argued that the late 1910s were periods of poor harvest and large-scale crop losses in the Lasta and Wag areas of northern Wollo, which gave rise to raidings and wars against the lowland pastoral people adjoining the two areas. The same author states that the 1920s ushered in a period of serious decline in agricultural production in Lasta. The decade of the 1940s seems to have been relatively free of acute food shortages, for many of the older of our respondents in Ambassel (those old enough to remember major events in the 1940s) recalled the hunger of the early 1950s (probably in 1952 or 1953) as the first major food crisis that they had personally experienced. Since then famine has occurred in Wollo at least once every ten years.

According to peasant informants in our awraja and surrounding areas, the years of famine in these localities were 1962/63 to 1964/65, 1973/74, 1977/78, and 1984/85. All peasants interviewed agreed that the last famine, which in some parts of the province was said to have started in 1982/83, was the worst disaster in living memory.

There is some documentary evidence about the famine of the 1960s, although the exact date of occurrence and the magnitude remain in contention. A recent work speaks of a "devastatingly severe famine" in Wag and Lasta awrajas (northern Wollo) in 1966, which also spread to other areas of the province, killing several thousand peasants in one district in Ambassel in the same year (Mesfin 1984: 37). A document prepared by the provincial administration office (Wollo Province Development Office 1968) reports of serious famine in the province in the years 1963 to 1966, with the worst crisis having occurred probably in 1964/65.[1] The document states that as a response to the disaster (which is said to have affected "various awrajas" the authorities in the province established a food reserve scheme with storage facilities to be built in strategically located areas. Each peasant household was expected to contribute 1 *gunna* (5 kg) of food grain to the scheme to be kept in the reserve and distributed in times of shortages. The authorities estimated that a total of 240,000 quintals of grain which would be collected in four years was sufficient to eliminate famine from Wollo. In 1966 the scheme had raised a couple of thousand quintals of food (ibid: 7 ff).

So much has been written about the famine of 1974 that there is no need to belabour the subject, and readers who wish to pursue it may consult Mesfin Wolde Mariam (1984) and the references in that book. However, certain aspects of the famine are worth looking at again for comparative purposes.

Future historians may very well wonder how a tragedy of this magnitude could be followed by an even greater tragedy in the short span of a decade, and how the lessons of the past could so quickly be forgotten or ignored. In both instances, the tragic events unfolded themselves in dra-

[1] All dates given in this document and by our informants were in the Ethiopian calendar, and this may account for the discrepancy, 1964/65 is 1957 in the Ethiopian calendar.

matic fashion affecting millions of peasants, and no one could honestly claim not to have been alerted in advance; and yet the response of authorities at the local and national levels was by large similar in each case.

The 1973/74 famine remains unique as an event in several ways, and the train of political events that were set in motion in its wake have radically altered the course of the country's history. In this sense, it may be seen as the *faim fatale* of modern Ethiopia. This was the first food crisis in the country to attract world attention and to draw international emergency support on a large scale. Some have even argued that the famine may have contributed to the sophistication of the international media in handling natural disasters and in enhancing its ability to shape world opinion (Harrison and Palmer 1986). High powered, organized relief operation by foreign governments and international volunteer agencies (or NGOs) is a new experience in Ethiopia, dating back to 1973, and since then the country has continued to be the focus of international relief efforts, and has had to live with the murky politics of international food aid.

One of the first acts of the military government which came into power following the famine was to establish a Commission of Inquiry to investigate why and how the famine occurred, and the Commission reported its findings in October 1975. Based on archival material and interviews of a large number of government officials the report showed that the northeast was the hardest hit region of the country, and Wollo was the epicenter of the famine. However, drought, and starvation were reported to have occurred, though on a smaller scale, in parts of Gondar and Gojjam provinces, and in the lowland areas of Harrarghe and Bale. Interestingly enough, the chief causes of crop failures and food shortages in Wollo were prolonged drought in the *qolla* awrajas, and frost and pest infestation in the *degga* awrajas of the province.

The most important revelation contained in the report has to do with the elaborate and coordinated efforts of state officials at the national and provincial levels to cover up the famine, to prevent relief assistance from reaching famine victims, and to deny starving peasants the chance of life. From high ranking officials in the Ministry of Interior to those in the provincial administration in Wollo, from the Ministry of Agriculture to that of Information, energetic measures were taken to hide the famine from the international and local media, and to prevent starving peasants from seeking assistance or petitioning the government. Among the sixty or so officials executed by the military authorities in November 1974 for various political offenses, at least four were directly implicated in the cover-up of the Wollo famine.

If the cover-up of the famine revealed the callousness of imperial bureaucracy, the pervasive corruption that thrived throughout the famine in Wollo showed the greed of local officialdom. It should be noted here that both the Commission's report and evidence from other official sources point out that there was surplus food in other regions of the country as well as in parts of Wollo at the time, and that the crisis would have been contained if the available resources of the country had been sufficiently mobilized for the purpose. The government's agricultural marketing agency, the Grain Corporation, had stocks of food in storage in Wollo in 1973 and 1974, but it did everything possible to obstruct the sale of this food to either the peasantry or to private charity organizations in these years; it persisted in its refusal even after higher authorities in Addis Ababa had ordered it to release food for sale to peasants and relief agencies (Mermarri Komissyon 1975: 86ff).

Similarly, local officials either banned charity organizations from carrying out relief work or made it difficult for them to purchase food for distribution. Private grain merchants who charged exorbitant prices greatly benefitted by the obstructive policies of local officialdom in both 1972 and 1973. Later, when the government was forced to act through pressure from the world media, food aid sent to the starving areas was not distributed to the needy but sold to private merchants; in some cases, relief food was left to rot in storage because the officials concerned did not bother to take delivery of or to transport it to where it was needed. Starving peasants fleeing from the famine zones were forced to pay what the Commission calls "*ye-kotae gibr*" (literaly, "hoof tax"), a tax to obtain a pass to travel through the famine affected areas. One particularly enterprising official, the administrator of a sub-district in Wore Himenu awraja charged fifty Ethiopian cents per head of cattle for the pass, and the money ended up in his pocket (ibid.: 51).

Some peasants identified the year 1977/78 as a famine year, although others said they did not experience any shortages in that year. Those who reported famine were those who lived in the Qolla (lowland) zone.

Famine is a shattering experience and it often leaves an indelible mark in the minds of the rural population. No peasant will ever forget this critical event if he himself has personally been involved, and the collective consciousness of a disaster is expressed in the way peasants talk of the event. The 1974 famine is known in many parts of Ambassel as *Wajja* from the name of a small town in north Wollo between Qobo and Alamata; it is believed that is where the disaster first began. The 1984/85 crisis is referred to by some as the "*time men ate grass*": many peasants had eaten wild plants to save themselves from a slow, agonizing death. One peasant we interviewed on the slopes overlooking the Gerado plains west of Dessie referred to the 1936/37 famine as the "Luqas of the Italian Occupation" from Luke, the name of one of the Gospels (see below for a discussion of this). Older peasants occasionally recount stories of past famines and similar tragedies to younger peasants to highten the latter's awareness, and to prepare them for future ones. Such stories also serve to reinforce respect for traditional values especially those having to do with survival in difficult conditions: *diligence, frugality, cooperation.*

Famine awareness also finds its way into folk literature, traditional sayings and proverbs. Wollo, and particularly Ambassel is noted for its folk songs and poetry which cover a wide variety of subjects including hunger, hunger-induced migration, peasant grievances against injustice, etc. Readers who may wish to pursue the subject may start with the Amharic novel about the Wollo famine by Berhanou Zerihun (1980). A fine song-poem about famine in the northeast commonly recited-sung by the *lalibella* or *hamina* (the Ethiopian version of the *griot*) probably at the turn of the century is treated by Berhanou Abebe (1970).

One of the questions we asked respondents in Ambassel was: do you believe famine is inevitable and will always occur? The majority of peasants were unsure, but 12 per cent said it was inevitable and 17 per cent said it was not, as shown in Table 14.

Table 14. *Will famine always occur?*

	Yes	No	Unsure	Other	Total
Number	31	44	178	2	255
%	12	17	70	–	100

Interestingly enough, peasants in Ambassel woreda which was relatively less affected by the recent famine or by the one ten years earlier were more "pessimistic" than peasants in the other two woredas; 21 per cent of them said famine will always occur while the comparable figure for peasants in Worebabo and Tehulederae woredas was 10 per cent and 8 per cent respectively. On the other hand a far higher proportion of peasants in the latter woredas were uncertain about the problem as opposed to those in the first woreda. It is our estimation that a good number of those who said they were unsure about the question gave that answer in order to avoid thinking about an experience which for them may have been a particularly painful one.

Older peasants were more pessimistic about future famines than younger ones. Of those who answered yes to the question more than half were in the 40 to 59 age bracket and about a third in the 50 to 59. Asked about the frequency of future famines 35 per cent of the respondents in the "pessimistic" category said they will occur with definite regularity, ranging from once every five to once every seven years.

Peasants, Moslem and Christian, consider drought and famine as an act of God. Only a small number of peasants tried to associate drought with climatic changes or changes in the natural environment. These were almost always younger and more literate and had been exposed to the media, especially to radio broadcasts. We detected some difference among the great majority who saw drought as divinely inspired. There were those who accepted famine as a manifestation of the wrath of God (*Ye-Igziabher gouta*), and a punishment for Man's misconduct and disregard of his moral duties. There were others who believed only the Supreme Being could cause the rains to fail, but considered that the divine purpose of the act was a mystery that will not be revealed to ordinary men, or that ordinary men like themselves were unable to fathom.

The explanation of social calamity as an act of a redeeming God is perhaps as old as religion itself, and has been the stuff of the apocalyptic visions of Ethiopian sages through the centuries. "Our lives are in the hands of God" was a frequent saying of peasants in Wollo at the time of field-work, but, sad though it is, God has shown his wrath more often than his mercy. Five terrible famines in the 19th century, six devastating ones in this, and more to come before the century is out: such has been the tragic history of the peasants of Wollo and the northeast. In the light of this grim experience it is understandable why the religious and the fatalistic emphasis looms large in peasant consciousness. However, the supernatural explanation of calamity frequently offered by peasants contains a subtle political meaning hidden in it. It expresses the despair of the people with existing social institutions, and their alienation from the matrix of power which defines their lives. Fatalism and other worldliness is born of hopelessness, resignation and withdrawal, but it also reflects disappointment and outrage at the reality of social arrangement. It may indeed be considered a form of spiritual protest.

When the full history of the 1984/85 disaster in Wollo is finally written it will show that famine had given sufficiently loud warnings to all who would listen that it intended to strike again. Some of our informants, peasants as well as extension agents, stated that in the lowlands of Worebabo and Tehulederae woredas the food crisis really began in 1982, when both the *belg* (spring) and *meher* (autumn) rains were severely inadequate. For most of the awraja as well as the province, however, the march of famine starts in earnest in 1983 when the main season rains in the autumn were short by between 26 and 60 per cent, the latter figure reflecting conditions in the lower elevations. There was a substantial loss of crops in the *qolla* and lower *woyna-degga* areas of Ambassel and Yejju awrajas, and in most areas of Raya Qobo, Wag and Lasta. In the west of the province, Wadla Delanta and Wore-Himenu were seriously affected in this year. In the Afar lowlands, which are permanently food deficit areas, the problem was not so much crop failures as the lack of sufficient moisture to enable grazing areas and watering points to sustain the large cattle population owned by the pastoralists.

In war-torn areas of northern Wollo, particularly Wag, and later Lasta, famine conditions begin to threaten the rural population as early as 1981. Partly because of repeated crop failures, and partly because of the fighting and civil disruption here, large numbers of peasants started to migrate to the roadside towns of Korem, Alamata and Qobo in 1982. The shelter population in Korem begins in this year with about 5,000 peasants, mostly from Wag, who arrive in the town hungry, exhausted and in poor health and refuse to return to their home areas. One eye witness report estimates that by early 1983 there were over 21,000 peasants in make-shift shelters in Korem alone ("Anonymous" 1983). One of the first feeding centres to be established in Wollo by international volunteer organizations is set up in this town in January 1983 to care especially for the young and the elderly but also for peasants of all ages who had fled war, destruction, and famine (for a general history of the famine see Hancock Gill). There was very little emergency relief assistance at this time, and RRC became involved in food distribution to the needy in earnest in the last months of 1984. A number of feeding and medical shelters were built in 1983 in several parts of the province by NGOs but the bulk of the victims of hunger went unnoticed, or uncared for. According to the records of the Dessie branch of MOA, about one-third of the rural population of the province was in need of food assistance by the end of 1983, a figure that may be a little inflated. On the other hand, the awrajas of Borena and Wore-Illu, and the western portion of the awraja of Dessie Zuria were relatively free of trouble. So too was the peasantry of western Ambassel, especially those in the upper *woyna-degga* and *degga* zones.

The crisis intensifies when the rains in the spring of 1984 fall far short of expected, leaving many peasants completely destitute. As if this was not bad enough, the autumn rains failed almost completely, bringing the inexorable and uncontested march of famine to its triumphant climax. Between 1983 and 1984 some thirty shelters and feeding centres were established in the province and run by about a dozen NGOs. Some of the main shelters which held large inmate populations were, however, put up in mid-1984 when hunger victims began to flock to the urban centres in exceptionally large numbers.[1]

The main shelters were built in the roadside towns along the Dessie–Megelae highway in the north, and the Dessie–Assab trunk road to the east. By October 1984, there were large shelters (sometimes more than one in a single town) in Qorem, Alamata, Qobo, Mersa (in Yejju) and Haiq along the north route, and in Millae and Bati along the east. A small feeding and medical centre had been operating since 1983 in Lalibella in Lasta awraja, and there was also an emergency centre at Harbu, just south of Kombolcha, which was set up in October 1984. There are some valid reasons why the shelters should have been set up in this manner—accessibility, proximity of peasantry to their home areas, large migrant population already in the towns concerned, etc.—nevertheless, there was also a determination on the part of local authorities to prevent the migrants from flocking into the capital and then treking to Addis Ababa via Dessie.

There are two other aspects of the problem we are dealing with worthy of note. The first is that mass migration of hungry and bedraggled peasants begins in northern Wollo in 1982 and reaches a high point by mid-1983; in southern and eastern Wollo, this phenomenon picks up

[1] For a blow-by-blow account of the march of famine and the international response, see Tom Franklin, "The Build up to the 1984 Ethiopian Famine". Unpublished report and part of a forthcoming book, Addis Ababa, U.N. Office of Emergency Operations, 12/8/86. For a critical review of the tardy and ineffectual response of domestic and foreign agencies see Cutler and Stephenson: 27–34.

momentum in mid-1984 and rises to an explosive level in August and September of that year. The second aspect is that local government agencies begin to stir themselves up and prepare for relief and emergency work after the end of October 1984, that is, a few days after the government formally announced the existence of a grave food crisis and issued guidelines for measures to be taken.[1] This comes out quite clearly in most of the documents prepared by the authorities in Wollo, and indeed November 1984 is the main reference point and cut-off date for all activities related to famine relief and resettlement here (see esp. documents prepared by NEEPR, NEERNDRC).

The death toll (1984–1985)

It is impossible to measure, or make a meaningful estimate of the death toll at any level of administration in the province in these two tragic years. There are two main reasons for this: first there were numerous "places of death," and most of them did not allow records to be taken, and second, no systematic records were made in those places where it was possible to do so due to a variety of factors. Death through starvation, exhaustion, and hunger-induced disease occurred: (a) in the home; (b) out in the open: on the road, or on rural footpaths where people, walking somewhere in search of help succumbed to death out of sheer exhaustion. People also died in church or mosque yards, and in public places. (c) Deaths occurred in the camps and feeding centres. There were often two sets of camp people, the "inmates", those who were inside, and the "out-mates", those who were not taken in by the camp authorities because there was no room, and who settled around or near the camps in make-shift shelters. Some camp authorities catered to the outsiders as well, and kept records of deaths, births, etc., of this population, others did not. (d) In collection centres while awaiting transportation to the resettlement areas.[2]

---

[1] A letter (undated) written probably in autumn 1984 to an official of MOA in Addis Ababa by MOA in Dessie reports that RRC had distributed a total of 73,445 quintals of food in the province in that year, a pittance given the magnitude of the demand for food.

[2] It is remarkable that published RRC documents do not provide figures on human mortality, and—this is deliberate—also make it impossible for anyone to pursue the subject.

The famine of 1984/85 may be considered the worst rural tragedy to have struck the country. The story of its magnitude has been dealt with elsewhere and there is no need to reproduce it here. It suffices to say that some 10 million peasants, i.e. about one-third of the rural population of the country were affected by drought, hunger, starvation and disease in these two fateful years. Partly due to this, and partly to the greater exposure of the tragedy in the world media, mortality estimates have varied far more widely than in any previous occasion. The issue is made more complex because large-scale deaths occurred caused not just by the famine *per se but also by the massive relocation programme* which the government embarked upon at the critical moment during the *famine*.

The least credible estimate is that given by some foreign journalists and writers according to whom one million deaths or more had occurred in the years between 1984 and 1985 (see, eg. King 1986: 36). This estimate has gained wide currency in the Western media. More informed observers on the other hand have offered a figure of half a million, but this too is quite satisfactory. "On the basis of the collective judgment of donor missions, relief officials, and representatives of NGOs working in various parts of Ethiopia," says Kurt Jansson, the head of the UN agency for Emergency Operations in Ethiopia, in his bland and accommodating book, "my very tentative and reluctantly arrived at view is that during 1984 and 1985, probably up to 500,000 people died as a result of the famine" (1987: 74; see also World Bank 1985).

Our own view is that the death toll may not exceed 400,000, of which 100,000 may be taken to be deaths that occurred in Wollo province alone. This refers to famine deaths only (i.e., not including deaths that occurred due to resettlement), and covers the period between the first quarter of 1984 when the death rate began to rise sharply, and the last quarter of 1985 when the crisis was sufficiently stabilized. Let us look at the problem more closely.

To avoid an over-extended discussion of the subject we shall restrict ourselves to some aspects of mortality in and around Ambassel awraja. First, a few words about the spread of the famine and the relief effort in 1985. Nature was just as relentless in this year as it was in 1984, and the *belg* rains were again inadequate shattering many more peasants' expectations. The relief effort was greatly accelerated in the last quarter of 1984; in November of that year over one million peasants were receiving emergency food aid in the province as a whole. In the first quarter of 1985, the number of aid recipients averaged about 720,000 a month and declined thereafter reaching a low of

324,000 in July when the weather conditions appeared to improve somewhat. However, there is a slight rise in the number again beginning in August; between this month and the end of October approximately 400,000 peasants continue to receive support each month. It should be noted that in many food distribution centres more people received aid than were entitled to, as a number of PAs, who were responsible for registering the needy for assistance, were often not averse to including those who were capable of supporting themselves.

In the first quarter of 1985, about 67 per cent of the rural population of Ambassel was registered as aid recipients but as noted earlier, the figure is higher than it would have been if far more stringent measures were employed to screen applicants for aid. Table 15 gives a breakdown of food recipients in the awraja's three woredas.

Table 15. *Registered food recipients in Ambassel by Woreda*

Woreda	Number of Recipients	% of Rural
Ambassel	83 845	70
Worebabo	59 835	71
Tehulederae[a]	100 326	63

Note: [a] Peasants in this woreda were also registered in distribution centres in the adjoining woredas in Qallu awraja.
Source: NEERNDRC, May 1985.

Let us look at the death toll in two shelters, one in Haiq, the awraja capital, and the other in Bati, just across the border in Qallu awraja where the majority of the inmates were Ambassel peasants. The Haiq shelter was a relatively small one and was administered by RRC in the awraja. The records of this office, which we investigated in October 1986, were neither consistent nor complete.

Table 16. *Death toll of Haiq RRC shelter in the period Sept. 1984 to August 1985, by age and sex*

0–6		6–15		16 & over		Total		Grand Total
M	F	M	F	M	F	M	F	
202	181	167	147	231	134	600	462	1062

Note: The age classification is that employed by RRC for food distribution purposes.
Source: RRC Records, Haiq (Ambassel).

There are two things about the data in Table 16 that should be noted. First, the death toll for males is considerably higher than for females in all age brackets. Second, the number of adult deaths, i.e. those in the age bracket 16 and over, is much higher in comparison to what conventional wisdom would lead one to expect.

Bati was the second largest emergency camp in Wollo after Qorem, and was run by the Ethiopian Red Cross. At its height it held an inmate population of over 33,000, a great number of whom came from Worebabo and Tehulederae woredas in Ambassel. Bati was started in October, but in the months prior to that, thousands of peasants had flocked to the town and were living in appalling conditions in make-shift shelters. It was in response to the urgent need to do something about the mounting problem confronting these helpless refugees, and the need also to avert a catastrophe which was imminent, that Red Cross authorities decided to establish a shelter and set up medical facilities there. The inmates were housed in 700 large tents, and about 385 medical staff were deployed (of which 86 per cent were foreign personnel) to provide health care to the population. Outside the camp were several thousand destitute peasants waiting to be admitted in the early months of the camp's operations. In this period also, the sexual composition of the camp was not overly uneven, but after the first quarter of 1985, there were far more women than men. The reason for this was that in March and April (the *belg* season) some rain had fallen and the men had left to see if they could work their land.

Table 17 gives monthly figures of the camp population including the death and birth rates. It is unfortunate that Red Cross records, from which the data is taken, do not often break down information by age, sex, or place of origin.

The highest number of deaths is recorded in the month of November 1984 when on the average 80 peasants died every day. The period from this month to February 1985 shows a high but declining death rate which continues to decrease until June when it rises to 465. By the time the camp was closed on November 7, 1985, the death rate hade been reduced to less than a dozen a month, although there were a total of 25,716 inmates still in the camp by the end of October.

The government's emergency resettlement programme was announced in October 1984, and a few weeks later officials in Wollo started the implementation process in the province. According to the policy guidelines a total of 300,000 peasant households, or about 1.5

million people were to be relocated in the western lowlands of the country, including the hot semi-arid areas of Mettema in Gonder, Mettekel in Gojjam, and Gambella in Illubabor provinces. Of the total number, about half, i.e. 150,000 households or 750,000 persons were to be from Wollo province. Local authorities here were not certain how such a large number of people could be collected, neither were they in a position to provide the necessary transport services and other logistics. Amid the haste and confusion it was decided to assign quotas to each awraja, and thus, more or less arbitrarily, Ambassel officials were given the task of preparing 12,000 households (or 72,000 peasants) for resettlement. The latter in turn assigned quotas to each of the three woredas of the awraja and instructed the woreda officials to collect the required number.

Table 17. *Bati shelter population basic data, 1984–1985*[a]

Month	Shelter Pop.	Deaths	Births
October (1984)	22 836	339	44
November	28 112	2407	80
December	23 163	1617	97
January (1985)	27 228	1110	121
February	22 463	776	84
March	23 813	346	118
April	17 121	185	65
May	23 440	185	49
June	24 552	465	47
July	8169	306	28
August	33 549	83	18
September	29 149	24	14
Total	–	7844	765

Source: Red Cross Records, Bati, Wollo, October 1986

[a] A series of large charts containing the above information and other basic data about the shelter population were hung in the Bati Red Cross office for several months and shown to senior government officials, international visitors, and ordinary guests. The charts were still there when we visited the office in October 1986.

In each woreda (except Ambassel)[1] there were collection centres where peasants recruited for resettlement from the kebbelaes were tem-

[1]In Ambassel woreda, Gishen and Tis'Abalima were used as collection centres but were abandoned after a short time, and peasants from the woreda were sent directly to Haiq.

porarily sheltered until transport was available. From here they were trucked to Haiq, Bistima and Gerba (all in the awraja), and Kombolocha in Qallu awraja to await transportation to the resettlement areas. In each of the primary and secondary staging points peasants spent several days and nights, often in appalling conditions, and many deaths occurred as a result (a discussion of the resettlement process is given in NEERNDRC May 1985: Ch. "Le"—the Amharic alphabet).

Table 18 gives a woreda breakdown of the number of peasants sent for settlement from Ambassel in the period November 1984 to April 1985.

Table 18. *Peasants sent for resettlement from Ambassel awraja*

Woreda	Number		
	M	F	Total
Ambassel	6116	5424	11 540
Worebabo	5141	4918	10 059
Tehulederae	8163	7110	15 273
Unidentified[a]	–	–	5959
Total			42 831

Note: [a] These peasants, who were sent in May 1988, were not identified by place of origin or sex.
Source: NEERNDRC May 1985:31.

The deaths that occurred in the primary and secondary collection centres are given in Table 19.

Table 19. *Deaths during resettlement in Ambassel awraja*

Primary Centres				Secondary Centres			
Woreda	M	F	T	Centres	M	F	T
Ambassel	–	–	–	Haiq	263	221	484
Worebabo	262	192	454	Bistima	140	118	258
Tehulederae	223	214	437	Gerba	83	75	158
Total	485	406	891	Total	486	414	900
Grand total		M: 971		F: 820		T: 1791	

Source: NEERNDRC May 1985:32, 44, 52.

A comparison of Tables 16 and 19 reveals one interesting fact: more males died, proportionally or in absolute terms, than females both in the shelters and during resettlement. This holds true for mortality in at least two other shelter camps in Wollo. More males than females died in the "out-shelter" famine population, and also in the shelter population in Qorem, except, in the latter case, for the period covering January to August 1985 according to information provided by Save the Children's Fund (records of Ministry of Health, Dessie). In Harbu, another high mortality shelter run by an Irish NGO (CONCERN), 16 per cent of the male but 12 per cent of the female shelter population died in 1985 (NEERNDRC, May 1985, QA: 28, 31). If this sexual distribution of mortality holds for other famine areas in Wollo and the northeast then we have a number of intriguing questions to answer: What are the reasons for this? Are women more resilient than men in the face of hunger or do women take advantage of their key position as family food managers? There is no question that more research is needed in this area, but in the meantime those with a tendency to see the exploitation of women in all famine experiences are advised to temper their judgements. (Further data about the same issue is given further on.)

One of the questions in our questionnaire was about family deaths among our respondents in the recent famine. Thirty per cent of the peasants in our sample said they had lost at least one member of the family in the famine; the distribution of this by woreda is: 20 per cent of respondents in Ambassel (with 13 deaths), 29 per cent (47 deaths) in Worebabo, and 39 per cent (61 deaths) in Tehulederae. The total number of family members lost was 125. Once again, and as shown in Table 20, more males are reported to have fallen victims of starvation than females.

Table 20. *Family members lost by respondents in Ambassel in the famine*

Hhds which lost[a]	No. of Hhds	Loss by Sex		
		M	F	T
1 Member	43	25	18	43
2 Members	23	24	22	46
3 Members	8	12	12	24
4 Members	3	6	6	12
Total	77	67	58	125

Note: [a] No deaths over 4 were given. A Red Cross Survey carried out in Ambassel and Qallu awrajas found that 51 % of respondents had lost at least one child in the famine (1986, Vol. III:81).

The real magnitude of the famine toll in Ambassel or in Wollo will probably never be known, but the investigation of the problem along the lines indicated here, and information gathered through oral and family histories in other parts of the province or the region may yield valuable results.

III SURVIVAL: COMMUNITY AND COOPERATION

7. The Community in Distress

What is remarkable about the last two famines in this country is that in both cases the peasant community was abandoned by the outside world and left to its own devices when disaster struck—i.e. at the very moment when it needed the largest number of friends, and deserved the greatest amount of assistance. It is as if the rural world had contracted some terrible disease, and everyone had decided to shun it, and to enter a conspiracy of silence lest people of delicate constitutions be shocked and offended by the news. The darkest hour of any peasant society is the hour of death by starvation, and all over the northeast, in 1974 as 1984, thousands of men and women, young and old, succumbed to the worst form of death before assistance and relief reached the region. In both instances, when emergency aid finally arrived it was for many too late or too little.

Left to fend for themselves, the peasants of Wollo attempted to cope with the crisis by mobilizing the resources of their communities, and by utilizing their accumulated knowledge and experience to the best possible advantage. In both instances, the effort involved collaboration and co-operation as scarce resources were carefully managed, and the potentials of the surrounding environment thoughtfully investigated and put to use. In the rural world, resources are often communal or group-oriented, and knowledge and experience are in large measure collective property. It is true that relief and emergency aid was responsible for saving the lives of countless peasants, but it is equally true that indigenous survival techniques and the collective efforts of the peasants themselves were instrumental in saving a greater number of people from death.

Contrary to certain accepted views, which mercifully are now becoming more and more rare, a peasant community does not just give up and wait for death when confronted with food crisis. As was argued at the outset of this study, the threat of imminent famine triggers a heightened awareness, and a spate of defensive activity within peasant communities. Everything is done to anticipate, prepare for, withstand, and finally withdraw from the ambit of the crisis.

A simple famine scenario looks something like this: the failure of the rains in spring means no harvest to tide peasants over to the next season. By the end of May peasants' food stock has run out or is dangerously low, and they make ends meet in various ways, including asset disposal. The failure of the main rains in June and July begins to trigger a trickle of migration from the rural to the urban areas, and this is then turned into a torrential flood in September and October when the last hope of the peasantry to salvage something out of the extreme situation is dashed with the failure of the rains in the month of August.

This is of course a simplified scenario, and does not take into account a variety of factors, such as civil conflict, sporadic precipitation, positive or negative response of the market due to outside influences etc., which may accelerate or reduce the tempo of the final drama, which is death and dispersal. As we shall see further down, this scheme does not quite fit the march of events in Ambassel in 1984–85, but we shall retain it for analytical purposes.

We shall identify three phases of famine response in rural society: crisis anticipation, crisis management, and exhaustion and dispersal. In the pages that follow we shall discuss each of these phases taking the experience of the Ambassel peasantry as case material.

Crisis anticipation

By crisis anticipation we mean all activities of a predictive and defensive nature employed to foresee the probable course of events in the immediate future, and which help a peasant household prepare alternative resource management strategies in order to forestall or minimize danger. We are referring, in other words, to indigenous systems of disaster forecasting and disaster preparedness. These systems, based as they are on long years of peasant experience, on traditional conceptions of ecology and environmental change (or "Folk ecology" as one writer calls them—Richards 1975), and on indigenous agronomic know-how, may be considered "unscientific" or "pre-scientific", nevertheless, they form an important part of peasant survival strategies and deserve to be examined. The effectiveness or ineffectiveness of these systems is of course difficult to measure with any degree of accuracy, and the charge of "unscientificism" applies not so much to the purpose or even methodology of the systems as to the inputs and parameters employed to measure probabilities and to assess ecological behaviour.

A comparison of modern early warning techniques with traditional ones will show that despite the "scientificness" of the former and its use of advanced technology, and "advanced" methods, its predictive value is not overly superior to that of the latter. Traditional forecasting in fact employs some of the same indices of disaster assessment as modern forescasting, and may, in certain instances, be better at sensing imminent danger than its modern counterpart. Let us digress here and discuss briefly modern food crisis forecasting to illustrate our point and to support our argument. The discussion that follows draws upon and is in part a critique of the following works which have dealt with the subject: Seaman and Holt (1974), National Research Council (1979), Cutler (1984), Holt and Cutler (1984), and Cutler and Stephenson (1984).

"The state of food emergency preparedness in Ethiopia", says a recent document, "is extremely poor" (Cutler and Stephenson: 45), but the same document had in an earlier chapter praised the country's capabilities in this area: "Ethiopia is probably the best-equipped country in Africa in terms of food emergency preparedness" (ibid.: 2). The Ethiopian Relief and Rehabilitation Commission's (RRC) early warning system has been in use since the latter part of the 1970s, when it was adopted at the recommendation of specialists working for donor agencies and western governments. The system, as defined in several RRC documents, consists for the most part of evaluation of several indicators believed to be harbingers of crisis, the most important of which are weather and crop conditions, market behaviour, and rural migration; in collaboration with ENI, RRC also occasionally assesses the nutritional status of a population suspected of being in imminent danger. Its main source of information are the Ethiopian Meteorological Service (EMS) for weather behaviour, and CSO and MOA for crop conditions. Additionally its own surveillance team may take visits to the areas concerned to carry out on the spot assessment of general conditions, including the condition of the agricultural market. (For the official view of RRC's early warning system see RRC, Dawit 1984, Ahmed Ali 1986.) In a number of cases the surveillance team is not composed of specialists and expertise relevant to the problems to be investigated. The system looks simple and effective on paper, but the actual process of emergency forecasting employed by the agency is far more bureaucratic and involved as we found out in the field in Wollo.

According to the information provided us by the chief officer of the Ambassel awraja RRC, the standard process by which information about food shortages or serious drought in a particular locality is transmitted to concerned authorities has not changed significantly from the days of the Old Regime (see Mesfin 1984 for the old practices). Neither the awraja nor the provincial office has capabilities for monitoring agricultural, marketing and health conditions in the rural areas. Neither carries out grain or livestock surveys, nor crop production assessments. This is how food shortages or drought in an affected area filter out to higher authorities.

Local peasants or PA leaders (sometimes lower level extension agents) report the problem to the woreda administration office. If the administrator has reason to suspect that there is indeed a problem, he sends out a team (occasionally he may himself lead a team) to investigate. If the team is convinced there is indeed a food crisis, it prepares a report of its findings and sends it to the awraja disaster relief committee (until recently such a report was sent to the awraja administration). This committee is made up of the local party chief, and senior government officials of the awraja, including the chief officer of the local RRC. The committee conveys its decisions in writing to the RRC office which in turn transmits this report and its assessments to the provincial RRC office. When the information finally reaches the head office in Addis Ababa via the provincial RRC, one of two things may happen: a) the RRC head office may decide to send a surveillance team to the locality in question and a more in-depth assessment of the problem is made; or b) the report may be filed away and the slow process of information flow comes to an end; this may be several months after the process was set in motion by peasants or PA leaders at the local level, and in the interval people and farm animals may have begun to die. Needless to say, this system of disaster assessment and forecasting is slow, cumbersome and overly bureaucratic.

The standard approach to food shortage evaluation relies primarily on the results of the assessment of the following four major indicators: crop production, weather and weather impact, market behaviour and nutritional status assessment—this last being mainly anthropometry (see Seaman and Holt, ENI 1976, SCF 1986). As noted earlier RRC often includes crisis-induced rural migration as an additional variable. These may or may not be combined with environmental information acquired from satellite imagery and remote sensing (see National

Research Council 1979 for that).[1]

However, quite often it is not so much the choice of indicators that is important as the timing of the assessment itself. As often is the case in RRC's experience famine has already begun by the time early warning assessments are carried out and the results evaluated. Indeed, the results of on-site investigations and other relevant research in this country—and we believe the same may be true of other countries in Africa—take many months to prepare and present in intelligible form (for the eastern African experience see FAO 1984). Moreover, the crop and weather impact assessment can only make sense if damage has already been done; it does not, in other words, involve "forecasting" in the proper sense of the term.

The question: what will be the course of events in the coming months? In its predictive sense is not the central question in the kind of modern early warning system employed by RRC? Rather, the main purpose in the system is to determine whether or not a condition of abnormality already in evidence will get better or worse, and whether or not the magnitude of the abnormality in existence at the time of the investigation is such as to lead one to conclude that the area in question is suffering from famine.

One can also raise questions regarding the reliability of the data collected on some of the indicators noted above, especially about market behaviour and nutritional status assessment.

We have argued in a previous chapter that rural price movements, specially in what we have termed area or regional markets are reliable indicators only of a condition which is already evident to the casual observer. The experience of some other countries has also shown that the market may behave badly and may in fact send the wrong signals to outside observers and disaster forecasters, as has happened in Bangladesh on several occasions (Ravallion 1985). The reliability of anthropometric evidence must also be questioned, particularly if: a) it is carried out on a small sample of the child population of a given area, and b) if it is done hurriedly and not over a sufficiently long period of time.

Even by Ethiopian standards rural children suffer from endemic malnutrition far more than urban children, and the rate of mortality among the under-five year olds is much higher in the rural areas than in

[1] One work recommends "Length of the Growing Period" as an early warning tool; see Henricksen and Durkin 1985.

the urban (ENI 1986: 12ff). Since child malnutrition is not just chronic but endemic the nature of it changes from one agroecology to another. Thus the malnutrition found among pastoralist children is different from that found among children in highland agricultural communities. Some specialists on the subject have in fact strongly questioned the validity of anthropometry and energy intake measurement in the Third World. Human performance and energy requirements, one specialist has noted, change considerably in different ecologies due to longterm genetic adaptation (Srinivasa 1983), and the measurement techniques employed in Africa and Asia were actually designed for the needs of Western societies.

Finally whether or not the indicators themselves are properly evaluated depends on the evaluators themselves, and there may on occasions be an element of subjective bias on the part of assessors and concerned officials. Consider the following example. In their assessment of the Ogaden area of the Harar province in 1974, Seaman and Holt (ibid.: 88ff) discovered that five out of the six indices they had used to measure famine conditions showed high abnormality. They point out that mortality among the young was high, livestock losses were unsually heavy, grain prices had climbed up steeply while cattle prices were low or stable, there was a high rate of "rural" out-migration, and an unusually large number of people were in debt. On the positive side—the only one of the six indices to show a positive correlation—they found that there was no serious malnutrition among children in the area who were given anthropometric tests. The final verdict of the authors was that despite all that, the area was not suffering from famine.

Indigenous peasant early warning techniques concern themselves also with crop conditions, weather and environmental changes, and rural market behaviour; health status assessment is the only area which lies outside the bounds of traditional forecasting. Peasants follow closely agricultural conditions not only in their immediate vicinity but in distant areas as well, particularly in those which maintain traditional links with their own communities. Information on this and similar problems is often obtained through the market system as discussed earlier. But there are also other sources of information, namely rural travellers, people going on pilgrimages, and early victims of famine who are often marginalized peasants, and who migrate to other areas in search of assistance or employment. The same keen interest is held by peasants with reference to market behaviour and price fluctuations, on the one hand, and weather behaviour on the other. We shall discuss the latter problem further on.

In brief, the main point we wish to make in this connection is the following: although the methods are rather crude, the purpose of traditional early warning, and the indices of measurement employed are by and large similar to modern early warning practices. Furthermore, the practices in both instances are primarily defensive, rather than predictive in nature. In the former case they are employed to assist peasants to rationally manage their scarce resources to forestall danger.

By predictive practices we mean those endeavours, physical or mental, which are employed to acquire advanced knowledge of events or developments before they actually take place; these enable a rural household to prepare appropriate measures to deal with the expected eventualities. Weather forecasting is a case in point. The strength of any disaster forecasting system—traditional or modern—lies in its ability to acquire advanced knowledge, and on the level of accuracy of the knowledge so acquired. Traditional forecasting systems—in common with modern ones—attempt to unravel the future, the better to understand and prepare for it, with the aid of rainfall and climatic behaviour assessment, traditional ecological knowledge, and traditional beliefs including shamanism and divination.

"Farm in the *meher* (autumn) season with the lessons of the *belg* (spring) season in mind" is a popular saying in the rural areas, and many of our respondents in Ambassel readily referred to it when asked about traditional forecasting practices employed in their communities. Many peasants consider that the behaviour of the rains in the spring is a portent of things to come later in the year, and the experienced farmer files away in the back of his mind any peculiarities he or his friends may have observed in the early part of the year. The general understanding is that any irregularities in the *belg* season will often have effects in the *meher* season.

As was observed in an earlier chapter the *belg* season is important to all peasants for the following reasons: (a) For those who regularly practice bimodal agriculture it is a means of making up their food deficits. (b) For those who do not habitually farm in the spring the rains are welcome because they soften the soil, and their plots are turned into additional grazing grounds for their livestock. All peasants are thus concerned about the behaviour of the rains and the weather in this season, although some are more so than others. Peasant apprehension is aroused if the *belg* rains are irregular or below average. If the rains are short in this season peasants fear that the autumn rains will be short, or their timing may be badly affected, i.e. they may begin late and end early. Too much spring rain is believed to be followed by below average rain in the autumn, perhaps beginning normally but ending early. Most of our respondents were of the opinion that two rain failures in a year are exceptional, and the absence of precipitation in the spring does not often lead to similar results in the autumn but usually too reduced or untimely precipitation.

Readers who may think this is merely peasant superstition and therefore nonsensical should pause before they whip out their pens and damn the practice for posterity. The basis of the causal connection between the spring and autumn seasons observed by peasants is not fictitious but the product of the experience of generations of farmers, and there is reason to believe that this collective knowledge has stood rural producers in good stead.

The scientific evidence does not fully support traditional peasant climatology, neither does it fully disprove it. Consider the rainfall data for Haiq station in Ambassel for the twenty year period between 1963 and 1982 (see Annex 1). The data shows that irregular *belg* rains were followed by irregular *meher* rains in twelve of the twenty years upto 1982. We took the months of March, April and May on the one hand, and July, August and September on the other as the best and the least variable months for *belg* and *meher* seasons respectively. This gives a rate of coincidence—or should we say of accuracy—of 60 per cent. Four of the twelve irregular seasons were those where too much *belg* rains were followed by too little *meher* rains; another four irregular seasons showed too little *belg* rains followed by too little *meher* rains, and a final four when poorly timed *belg* rains led to an irregular *meher* in which there were heavy rains in the early months but below normal rains subsequently.[1]

The importance of the practice should not of course be overemphasized, but at the same time the rationality of peasant thinking should also be clearly understood. Irregular or short rains in the *belg* season will certainly lead to poor results for those peasants who have planted, but a peasant in this position may have enough harvest to recoup his seeds, and sufficient crop residue to feed his livestock until the *meher* rains. The

[1] The rainfall data was provided to us courtesy of Daniel Gemechu Geography Department, Addis Ababa University, see References for his paper on the subject.

main reasons why peasants are concerned about the behaviour of the *belg* rains, and attempt to predict the likely behaviour of the *meher* season are the following: (a) An unexpected irregular *meher* will mean that the timing of a peasant's farm-work may have to be changed. The peasant may have to decide to plough early, or late, or to prepare for dry-season seeding, as the case may be. (b) A peasant's selection of the appropriate crop mix will also be affected in these circumstances. If shorter rains are expected in the autumn peasants may decide to forego planting small-seed crops, which require too many ploughings and too much moisture, and opt for crops with low moisture requirements and more drought tolerance. (c) A peasant will have to decide what kind of seed activities to plant. If serious irregularities are anticipated fast-maturing crops will be preferred to normal ones.

The decisions that are taken in these circumstances will have a bearing on a peasant household's resources and the mangagement of these resources. While all peasants attempt to be as self-sufficient as possible, most peasants will often be in deficit in some crops and some seed varieties. These will then have to be purchased from the market, a decision which may involve selling small stock such as sheep, goats or poultry. The family (often at the initiative of the women in the house) may also decide to retrench until the behaviour of the autumn season is determined, and the expected harvest is considered to be normal or higher than normal. For some peasants, particularly those with depleted flocks, it may involve purchasing young animals, perhaps ewes and kids, which may be sold later if the autumn harvest turns out to be below normal.

Most peasants possess a stock of crude environmental indicators which they employ to evaluate changes in the weather on a day to day basis. Wind movements, cloud behaviour, temperature levels and levels of humidity, and on occasions the bodily reactions of farm animals are used to try to determine changes in the weather. Needless to say, older and more experienced peasants are more listened to than younger ones in this matter. This is of course short-term forecasting, and its value is limited as it does not lead to significant changes in a peasant's work or resource management plans.

Magic, divination and ritual

Mystery, mysticism and ritual characterize the traditional "science" of peasant society, a science which is put to its severest test in times of crisis, and which often fails when it is most needed. Like many other people who live in a pre-scientific age, Ethiopian peasants also wish to unravel the mysteries of the universe, but only if this has direct or indirect utilitarian values, and if the results are of benefit to themselves. The indigenous early warning system to be discussed under this rubric is partly predictive and partly defensive in content, and involves the interventions of the rural seer, and the collective involvement of peasant communities in religious, or religious-inspired ritual. It is also based in part on semi-occult divination based on what we wish to call arcane *seasonology*.

First, a word about the condition of the seer and medicine man in rural society. Petty and irresponsible officials in local government and political circles have made the life of traditional mystics, spirit men and healers intolerable in rural areas since the revolution of 1975. The homes of such people have on several occasions been raided or vandalized, and their contents frequently looted by gangs of zealots; these zealots have at times been assisted or given tacit encouragement by local law enforcement authorities. The individuals themselves have been beaten and thrown in jail for practicing their art, although there is no law prohibiting traditional practices of a mystical or medicinal nature. Occasionally, rural mystics have been ridiculed in public, denounced by political activists as charlatans and counter-revolutionaries. A PA leader in Worebabo informed us that about a year before the famine, something resembling a witch hunt had taken place in the woreda, and a few medicine men and diviners had been forced to leave the area; he did not know where they went to. Often enough peasants and the public in general have been warned by officials not to listen to or consult traditional specialists. The net result of the harassment campaign has been to drive these individuals and their specialist practices underground, making it difficult at present to collect information on the subject.[1]

The "men of knowledge" (as they are called in rural areas) themselves are now very reluctant to talk about their craft, and ordinary peasants will quickly deny any knowledge about specialists in his/her

[1] Some of the above information was provided by my colleague, Makonnen Bishaw (Sociology Department, Addis Ababa University) who is completing a dissertation on traditional medicine. The rest was obtained by us through interviews of peasants and peasant leaders in Ambassel.

community. The exception to this rule involves herbalhealers, and medical "repair men or women", i.e. those who treat sprains and bone fractures, do circumcision, treat bodily swellings and similar ailments by drawing blood, etc. As a general rule visiting a sorcerer or man of occult knowledge raises suspicions in rural communities as such visits are often interpreted as attempts by someone to harm another through the agency of malevolent spirits, or take undue advantage of certain opportunities.

We were able to find only three "men of knowledge" in our awraja, one in Ambassel woreda, and the other two in Worebabo. Two of them, however, refused to talk to us about their specialities, one outright, the other claiming he was only a herbalist and knew nothing about forecasting or divination. The third, a specialist living not too far from Bistima, the capital of Worebabo woreda, was briefly interviewed by my field assistant. According to this individual, the honest seer is so often confused with the *gallicha* that he frequently abjures his specialist knowledge and beliefs and becomes a simple herbalist or medicine man. The *gallicha* is related to, but not the same as the Oromo ritual institution called *gallu*; a *gallicha* is believed to practice spirit possession and trance. He is often considered to be engaged in anti-social and amoral behaviour and to practice sorcery for his own benefits, or against persons he disapproves of. He is generally feared and resented by the public. Much of the information provided below comes from interviews with peasants and PA leaders who felt "enlightened" enough to discuss what they knew and had heard about rural seers and diviners.

It appears that the "rain-maker", i.e. the diviner who either predicts rain or causes rain to fall is a rare, if not extinct being in this country; we certainly did not find or hear about one in Wollo. According to a traditional scholar, who is a consultant at the Institute of Ethiopian Studies (IES) here at the University, and a specialist in traditional medicine and the occult, there was to his knowledge only one reputed rain-maker in this country since the latter half of the century. This diviner, however, died in the mid-1960s taking all the secrets of his craft with him. Since then no "man of Knowledge" has appeared and demonstrated his skills at rain-making or rain-forecasting.[1]

[1] Interviews with Ato Ahadu Ayehu, Rere Ms. Division, IES, 10 December 1986. Ato Ahadu stated that this diviner was so highly reputed for his skills which he had demonstrated on numerous occasions, that he was invited to the Palace.

This is however contradicted by the ethnographic literature which speaks of rain-making as distinct specialities of magico-religious leaders in several ethnic communities in Ethiopia. Among the Oromo, there is the *malima*, i.e. one who brings rain (rain-maker), and the *chemsitu*, one who brings drought, or drives away the rain; this latter is often physically abused by the public if there is prolonged drought in a community. Both of these are specializations of individual *gallus* or ritual experts (Knutsson 1967). Among the Gurage, the same person acts as the shaman, medicine man as well as rain-maker (Shack 1971). We suspect this may also be true of the seer in Wollo. According to Ato Ahadu, the rural diviner nowadays has abilities similar to the *chemsitu*, i.e. of stopping rains, frost and hailstorms, and occasionally of providing antidotes against locust and pest infestation. Ato Ahadu, it should be noted, is knowledgeable only about the occult practices of the highland Christian people of the country.

The ethnographic literature on rain-makers and rain-making is however disappointing because the subject is not adequately covered, and the fine distinctions between rain-making and other ritual specializations is not often clearly brought out. Tippet (1970: 178) who studied the peoples of southwest Ethiopia reports that many ethnic groups here have specialists whose duties include rain-making. He states that the rain-maker among the Oromo of western Wollega is called the *irressa*, while Huntingford (1955: 77) argues that the *irressa* is the ritual expert in communal rain-making ceremonies in eastern Wollega where the rain-maker proper is called *camsitu*. Among pastoralists like the Afar and the Somal, rain-making may in the past have been one of the duties of the tribal or clan chief, but nowadays the task is performed collectively by means of a communal ceremony officiated by the tribal chief and assisted by the holy men of the community (Cossins 1972: 15).

The most extensive discussion of divination and rain-making appears in Trimingham (1965: 262–69), who argues that Islamized cultures in central and southern Ethiopia often contain beliefs that rains can be obtained or stayed by magico-religious experts. The same may be said of Christian cultures in the highlands. It should be noted that both Islam and Christianity super-imposed themselves on earlier, pagan belief systems, some of which they absorbed, and divination practices may have their roots in the people's pagan past. Rain divination, which includes the practices of haruspication, calendar prediction, and "animal reading" is employed for grazing strategies, long distance travel, rain and weather forecasting, and to predict sickness, epidemics, and natural calamities. According to Trimingham, the cultural groups which have strong beliefs in rain-makers (and rain-stoppers) are the Beni Amer and the Nera in western Eritrea, the Afar, Somal and the Oromo, and the Sidama and Hadiya in southern Ethiopia. Among the Christians of the northeast, the *debtera* (the unordained servant of the church who functions as a chorister) is often thought to have the power of making or stopping rains. In both Moslem and Christian practices, environmental diviners are almost always male.

Trimingham calls the rain-making ceremony among the Moslem Oromo in northeast Ethiopia the *raya*. The ritual always involves sacrificing an animal, usually a bull, a sheep or goat. Very often the ritual expert carries out the sacrificial rites on top of a mountain held sacred by the community; this is believed to bring the officiate nearer to God (Bartels 1983). Trimingham points out that certain animals are held sacred by certain communities, and divinations are performed based on the behaviour of the animals on certain periods in the year. The lion in the past was sacred to the Oromo and warriors were not allowed to kill it; similarly, the crocodile is held sacred by the Omotic people who live around the Chamo-Abbaya lakes in Gamo Goffa, southern Ethiopia. Even today, the Aderae of the walled city of Harar believe the hyena has magical powers and "read" its behaviour for omens and revelations on the occasions of certain holidays. Christian highlanders in the northeast however fear and despise the hyena, and often associate it with the evil eye or malevolent spirits.

However, the *malima* or the specialist with similar powers, i.e. rain-making and rain-forecasting, may have disappeared from many sedentary highland cultures since the end of the 1950s when the frequency of drought and problems associated with poor harvest increases relatively sharply. One suspects that the rain-maker will be assured of a sympathetic hearing among the public, especially the concerned public, in times when drought and famine are rare occurrences.

Those who may be tempted to argue that social crisis such as hunger and mass privation triggers a heightened awareness of "traditional science", and an intensified activity by specialists in this area, will be discouraged by the results of our inquiries in Wollo. Admittedly, we spent only a short time looking for specialists, and digging up information about the subject, nevertheless, the impression we came back with was that neither the Shaman not the diviner rose to the occasion, or were particularly sought after by the peasantry in the crisis years of 1983 to 1985. It is of course possible that our impressions may have been the result of the dearth of reliable information on the subject and the practitioners, and there is thus a need for further investigation before definitive conclusions can be drawn.

Nearly a dozen peasants in Ambassel and Worebabo woredas said they knew or had heard about diviners active somewhere in the awraja, but all but one said they had not themselves consulted them during the food crisis; the one who did was told by the diviner that the troubles brewing in early 1984, when the peasant went to see him, would continue for some time to come.

From those informants who were willing to talk about the subject we were able to gather that rural "futurologists", if we may call them that, who engage in environmental forecasting come mainly in two groups: those that divine the future from omens, environmental and natural signs or happenings, and those who may be called "seers" and receive their knowledge, as it were, through dreams, and periodic flashes of clairvoyance. In both cases, "divining" consists often of "seeing" in advance the coming of hard times and difficult conditions.

There were, we were told, three kinds of diviners belonging to the first category, although we had reason to suspect that there may be more than that number. Several peasants reported having heard of men who had the ability to understand the language of birds and to interpret the messages that one group of birds send to another. At certain critical moments these messages are said to contain forecasts about the environment and especially about the behaviour of the rains; these messages are intercepted by the diviner and revealed to those who come to consult him. One may well wonder if there is some scientific basis for "avian forecasting", since birds habitually move from one eco-system to another in response to or expectation of environmental changes. The specialist in avian studies may find it fruitful to investigate just what secrets the diviner of avian wisdom possesses.

It is worth noting that *ornithomancy*, that is divination based on the interpretation of the flight and behaviour of birds, is not an unknown practice in this country. Tippett (:170–71) found that some diviners in Wollega often combined this practice with haruspication to predict environmental changes and climatic behaviour. Huntingford (:77) likewise states that the songs of the sparrow are used to forecast changes in the weather among the Oromo in western Ethiopia.

A second group of "futurologists" are said to interpret subtle environmental changes on the evidence of bodily reactions of animals and specific types of surface cattle disease. Livestock, especially cattle and equines are said to become afflicted with certain mysterious ailments which are considered to be omens of hard times to come. This is not really divination but empirical knowledge accumulated through long years of experience mixed with anxiety or apprehension. Cossins (1974) has discussed several types of animal disease associated with drought and reduced levels of precipitation in the northeast; many of these diseases have been given specific local names by peasants in Wollo as well as Tigrai and north Shoa. Certain surface diseases are known to be fatal and occur in cattle at critical moments in a drought cycle. The shortage of pasture also drives cattle and equines to graze close to the earth, and this may lead to diseases affecting the mouth, teeth and throat.

A third group of "futurologists" are those who are said to possess the ability to "read the clouds" as one of our informants, a PA leader in Wichalae, put it. On occasions such men may not be diviners but community elders who are respected for their sagacity and moral probity. The secrets of this source of wisdom were not revealed to us but from what little information we were able to gather in the field, and from interviews with Ato Ahadu, serious drought is said to be imminent and famine inevitable if on certain critical days in the year the clouds appear in strange and unusual colours. Thick, heavy and unmoving clouds are a portent of bad times to come, whereas clouds specially dark red in colour—the colour of blood—are taken as sure signs of famine. The days when these "reading of the clouds" are done vary from place to place but the days following the Ethiopian New Year (11/12 September) or those following the Ethiopian Easter (18/19 April) are the most common ones.

It is also believed that some diviners in this group may also have the ability to "read the heavens". Such specialists are believed to derive their knowledge of things to come from their "reading" of the arrangement of the celestial bodies; such "readings" are carried out on specific days of the year similar to the practice discussed in the previous paragraph. Drought and hardship are said to be imminent if the configuration of the stars at night is unusual, if strange-shaped and giant stars appear in the skies, and if the moon assumes unusual and unhealthy colours. This form of divination is probably derived from the ancient

Ethiopian book of "futurology", *Fikarae Yesus*, which predicts the coming of the end of the world and warns the faithful to prepare for it. Some of the contents of this book are familiar to Church-influenced diviners, priests, and church-educated individuals in the rural areas, the coming of Doomsday is revealed to mankind by means of many heavenly signals, of which the appearance of giant meteors, or "stars with tails", as the book calls them, is the most significant. When this occurs, there will be "hunger and privation; men will consume human flesh, and eat grass like animals" (ibid: 61).

Astromancy, i.e. star divination, is not unique to the people of northeast Ethiopia but is also found among cultures in the south and southwest. Hallpike, for example, found the practice among the Konso of Gamo Goffa where it is known as "looking up at God" (1972: 176–78). Diviners here interpret the disposition of the stars and celestial bodies and make predictions about weather behaviour, wars, epidemics and the like.

The seer whose knowledge comes from dreams or through clairvoyance is an enigmatic character, and information about his craft, or of the stimuli for his visions are very difficult to collect. From what we were able to gather from a few peasant informants in Worebabo woreda, and from the brief conversation my field assistant had with the diviner in the same woreda, it appears: (a) That the seer, who also frequently doubles up as a medicine man, and a diviner of mysteries having to do with other aspects of life, becomes concerned if he gets visions which he interprets as serious environmental disturbances leading to drought and hardship. We were unable to determine whether or not trance or spirit possession was preceded by or accompanied the visionary experience; if the experience was obtained in dreams then most probably possession was not involved. If the visions are repeated several times he may be compelled to transmit the message to others and through them to the community. (b) He is not consulted on such matters like other specialists but reveals his visions and their meanings to relatives and trusted friends, who in turn disseminate the message to others in the community. Thus the community is warned of an approaching disaster, but only indirectly through the seer's informal messengers. (c) The message of the seer is always a message of hardship and gloom. Such people characteristically are not reputed to have the ability to forecast environmental changes of a positive nature, leading for instance to favourable circumstances in weather and agriculture.

There are two ways in which the messages and divinations of the "men of knowledge" are disseminated and evaluated in the rural community. The first is individual consultation, the results of which are passed on by word of mouth to relatives and friends. The second involves discussing the forecasts at an informal gathering of peasants held usually in the evenings and outdoors.

The gatherings are often small, perhaps involving twenty or so peasants. On this occasion, respected elders, informed individuals, and persons who at times dabble in sorcery and the occult will try to analyze the meaning of the messages; the gathering may also discuss the lessons to be drawn from past practices, and may decide to seek further "knowledge" from other specialists.

We now turn to a religious-magical practice which has deep roots in Ethiopian rural tradition, and which has its origins probably in magico-mystical beliefs and ritual. Both Christians and Moslems have their own separate symbolism and observations, and these may vary in emphasis, detail, and occasionally in content from one region to another. We shall first treat the practices commonly observed in the Christian communities of our awraja, and this will be followed by the observances of the Moslem communities in the same awraja. We shall begin with the strictly religious practices, and those sanctioned and often carried out on the initiative and with the participation of the clergy.

Both clergy and laymen believe that natural disasters such as drought, flood, earthquakes, epidemics, etc., are the work of God, and have a divine purpose, although some say this purpose is not revealed to all earthly beings. Acts of propitiation, such as for example slaughtering an animal for sacrifice, or making oblations are discouraged by the priesthood, but many peasants carry out one or both deeds, particularly if animals have to be slaughtered to mark an occasion. The clergy and the devout, on the other hand, prefer and insist on special prayers and fasts. This is the only occasion when the priesthood in the rural areas play a role in famine survival.

Either at the urgings of the clergy, or on the initiative of the public a community will decide to hold prayers and to engage in fasts at the early or middle phase of a drought cycle. Unless the priests insist on a specific day, the observance normally begins on the first day of the new month, and is repeated for seven consecutive days. The main part of the ritual consists briefly of the following: all adult men assemble in the churchyard on the day designated for the prayer which usually begins in the morning. Those who, for health or other reasons, are unable to travel to the church, must stand in front of their homes in a supplicant manner and in communion with the people at church. (On some occasions the observance may be carried out in an open place, outside or some distance away from a church). After a short prayer by the priests, the assembled congregation chants in unison and rhythmically a short supplication: Oh, Christ forgive us! Save us! Craraiso![1] This is followed by reciting in a low voice the first three words (they are three in Amharic) twelve times in succession using the twelve lines of the fingers of the right hand for counting. Those standing outside their homes for the observance are supposed to go through the same ritual at the same time as the congregation.

The observance may take anywhere from three to four hours, and no food is taken either by the participants or other members of the community including children before and during the ritual. Many peasants may fast the whole day, taking food only in the evenings. Some purists believe that not only humans but animals too should be involved in fasting, and households may separate young calves from their mothers for this purpose. It should be noted that this religious practice is exclusively a male one, and women do not participate in the ritual although they are expected to fast with the rest of the community. The ritual may be repeated for several months depending on the inclinations of the community and its priesthood. According to one peasant in Ambassel woreda, who was an ex-priest, the prayers of the people in his local community have on several occasions been answered and the rains have returned.

This brings us to what we earlier called arcane *Seasonology*. The origins of this belief and practice are unknown, but there are certain aspects in it which resemble traditional Chinese seasonology. The practice involves in brief predicting what a season will entail before the season itself begins. There are four seasons in highland Ethiopian cosmic conception and these recur in cycles. Each season or year is named after one of the Gospels, and the succession of seasons follows the order of the Gospels. The year or season of Matteos[2] (Matthew) is followed by that of Marqos (Mark), which is succeeded by Luqas

[1] "Craraiso"—I was unable to find the meaning of the word.

[2] The term in Amharic is "Zemen'e Matteos"... the season of Matthew but it may sound better in English to say the "year" of Matthew.

(Luke), and Yohannes (John). Each season thus named is closely associated with certain natural occurrences, giving rise to certain political and social events and tendencies. These occurrences and events are not random but are interconnected as will be shown later.

The year of Matteos: A year of peace and stability
The year of Marqos: A difficult year for Kings (ruling authorities) due to unrest and rebellions.
The year of Luqas: A year of famine and pestilence.
The year of Yohannes: War, bloodshed and death.

The reader may note that of the four seasons, which succeed each other in cycles, only one is believed to be a good and prosperous season, which tells us something about the outlook of ancient Ethiopian sages to life and society. Their view of social life is gloomy, dark and brutish.

The current year in the Ethiopian calendar is 1979 (it began September 11, 1986, and will end on September 11, 1987), and 1979 is the year of Luqas. The famine years in Wollo in the Ethiopian calendar were from 1974 (1981/82), which was the year of Marqos to 1978 (1985/86) which was again the year of Marqos. The year 1975 (1982/83) was the year of Luqas, a year of famine; 1976 (1983/84) was the year of Yohannes, and 1987 (1984/85) the year of Matteos. See Table 21 for more details. Readers who wish to pursue the subject further are referred to the ancient book of seasonology called *Metsehaff'e Aboushak'r* (the Book of Aboushak'r) edited by Ze-Menfes Qidous Abraha (under Abraha in References).

A common saying among the Christian peasantry goes something like this: "be-Matteos irres, be-Marqos iffes" (farm in the year of Matthew, and gather in a bumper harvest in the year of Mark). Devout Christians trace this saying to the influence of Shamanism and pagan thought. Harvest time in the highlands falls in the new year, and Matteos, the year of peace and stability enables peasants to devote their full efforts and time to farming. The restlessness associated with the season of Marqos may be a consequence of a phenomenon known in the highlands as *tigab*, which may be translated as a condition of abundance accompanied by high levels of consumption which gives rise to unruliness and rebelliousness.[1] Civil discord (i.e. Marqos) is believed to lead

[1] Merid W. Aregay (1986: 120) suggests *tigab* should be considered as individual non-confirmity.

to large-scale destruction, and to the neglect of agriculture, which is why Marqos is succeeded by Luqas, the year of hunger and privation. It was a common practice until recently—as recently as the early 1970s (see Cossins 1972, for example)—for famine to lead to raids and full-scale wars of plunder against neighbouring communities, and occasionally to inter-regional fighting. The regional uprisings and wars in north Wollo discussed by McCann (1984) were sparked off in part by inter-communal violence brought on by food shortages. No wonder then that Luqas is followed by Yohannes, a year of war and aggression. The cosmological cycle comes full circle with Matteos, when as a result of the exhaustion of men after several years of rebellion, hunger and bloodshed, peace and stability return to society.

Ethiopian seasonology does not end with this, there is an addition and in part a variation to the system. This makes it rather complicated and confusing to the average person. The sub-system ascribes a meaning to each day of the week as it falls on the Ethiopian New Year (11 or 12 September). According to this belief three of the seven days of the week are associated with poor harvests and food shortages, two days with civil unrest, and another two with widespread disease, and animal deaths.

If the New Year falls on a Sunday the harvest in the year is expected to be poor, there will be widespread disease followed by famine. Trade, however, is thought to flourish. If the New Year falls on a Monday, it is feared that the year will be a year of civil conflict, a bad time for commerce. If it falls on a Tuesday, the season is said to be a peaceful one, the harvest will be good, and general prosperity is forecast. Wednesday is associated with widespread uprisings, poor harvests and flood shortages especially in the second half of the year. Thursday is believed to usher in a difficult year, with many women suffering from a mysterious throat disease, and farm animals dying in large numbers. If the New Year falls on a Friday, the year will be an uneventful year, but it will be generally good for commerce. Saturday is associated with serious natural disasters (including locust infestation on a large-scale), hunger and widespread sorrow and anguish.[1] The clergy and devout churchgoers frown upon this doctrine which they say smacks of paganism and demonology. The ancient books in which this and the simpler system of

[1] I am indebted to the Academy of Ethiopian Languages, Addis Ababa (written communication, November 1986), for helping me out with details of this doctrine, which also has other aspects not relevant to our subject.

seasonology are found are according to the Ethiopian Languages Academy not recognized as religious works by the Ethiopian Church.

Putting the two systems together we obtain the following picture about the province of Wollo during the last six years (Table 21).

Table 21. *Seasonology and divination (Wollo 1981/82–1986/87)*

Year	Season & Meaning	New Year Day & Meaning
1974 (1981/82)	Marqos: Civil Unrest	Fri.: Uneventful year
1975 (1982/83)	Luqas: Famine	Sat.: Famine; Ecol. disasters
1976 (1983/84)	Yohannes: War, death	Sun.: Pestilence, death, hunger
1977 (1984/85)	Matteos: Peace	Tue.[a]: Good harvest
1978 (1985/86)	Marqos: Unrest	Wed.: Unrest, hunger
1979 (1986/87)	Luqas: Famine	Thur.: Pestilence, animal deaths

Note: The years in brackets are in the Gregorian Calendar.
[a] It was a Leap Year.

Looking closely at the table one could say that it was only on one occasion, the year 1977 (1984/85), that this divination system forecast tendencies and developments which turned out to be the exact opposite. This is not bad for a traditional system, which like ancient numerology, was designed for basically a preliterate population, and meant to instill in the labouring people fear of the Supreme Being, respect for authority, and concern for the values of hard work and frugality.

The peasants we talked to in Wollo were not fully cognizant of the doctrines of seasonology. Some knew the names of the four seasons, others some of the days associated with death and hardship, still others much less. Asked if they knew anyone who was knowledgeable about the system and the interpretation of the seasons, most answered in the negative, but a fairly significant minority answered in the positive. The practice in many Christian communities is to hold discussions on the significance of the seasons, and the prospects for the coming year, at an informal gathering of peasants and in the presence of a knowledgeable person about the time of the New Year. A knowledgeable person may be an ex-priest, a traditional diviner, a peasant with some church education, or an elder known for his food sense and wisdom. On occasions, a priest may be present in the gathering and may carry on a discourse on the mysteries of nature, and may comment upon the seasonological doctrines.

It is interesting that the Sidama also employ calendar divination to forecast climatic behaviour (Hamer: Ch. 6). The Gabbra, the Oromo-speaking pastoralists who live on both sides of the Ethio-Kenyan border, practice a complicated form of seasonology. Their seasonal calendar is used by local diviners to make weather forecasts and to advise on grazing strategies (Kassam 1984).

In substance the system of traditional disaster response employed by Moslem communities is different, but the purpose, and the timing is very often identical with Christian tradition. There are also certain similarities in ritual, and this may be because both systems draw their symbols from the same source, namely paganist beliefs and values over which the two religions were superimposed when they spread to highland regions originally.

The strictly Orthodox religious response of Moslem peasants involves fasting and prayer which is often initiated by the *imam* or other community religious leaders. This fasting-prayer ceremony is known as *Istisgah*. On occasions, and in some parts of Wollo, the ceremony may be preceded by another ceremony involving youngsters. Youths (usually of pre-puberty age) from one or more communities assemble together and travel to different well-known shrines in the area chanting and singing. These youths often carry the *ge'tema* (bulrushes), or wear it in wreaths around their heads; the plant is a symbol of fertility and rejuvenation. The purpose of the ritual is to inform and to call on the populace to turn to God and to prayer. The actual religious ceremony begins with a fast which is made for three consecutive days. On the fourth day everyone assembles in a public place and special prayers are said. The ceremony may be repeated on different occasions if the populace feels a need for it.

The non-Orthodox, and more ritualistic tradition is more complicated and different. Unlike their Christian counterparts, Moslems do not employ traditional predictive techniques similar say to *seasonology* discussed above. Peasant communities hold what are known as *wodajjas* for a variety of purposes. A *wodajja* is a session involving all or most members of a community such as a *got* or *mender*. There is a *wodajja* to implore the gods to make the crops grow healthy, held in April in the *belg* season; a harvest *wodajja* is held in October, and in this peasants pray for their harvest to be good, and for help, against crop disease. The *wodajja* against drought or for rain may be held any time there is a need for it; we may call this the rain *wodajja* or season.

The call to the rain session is often made by the *qadi* (religious expert). This takes the form of informing and acquiring the consent of the community on the matter, usually at Friday prayers. This done, the leadership selects two or three messengers to send to a reputed "man of knowledge" who may reside not too far, or a good distance from the community. The messengers are instructed to consult with this man, and to bring back his advice as to where, how, and when they should hold the rain session. This man of knowledge may chastise the community for going astray, for not following the true moral path, and for disregarding the great saints and religious teachers. This, he may say, is the cause of the drought and famine. He will then give precise instructions.

The seer or man of knowledge in Moslem practice is believed to consult the spirits that are thought to control the weather, or at times the venerated dead *welli* (saint or religious teacher) of the community concerned may appear in his dreams; he advises the messengers what to do, how to offer oblations and conduct the sacrificial ritual. There is an element of necromancy here. The officiate is the *abegar* but the *abegar* may occasionally dabble in occult practices.

The rain *wodajja* may be held under a tree frequently used for public gatherings. If such a place had once been used by a revered religious teacher, sessions held here are thought to be good. It may be held alternately beside a river which is reputed to attract the *malaika*, i.e. benevolent spirits. Or, it may be held by the shrine of a well-known local *welli*. If the man of knowledge has not given any preference, the *wodajja* is often held in this third spot.

This may probably be a heritage of pagan beliefs but the "shola" or "warka" tree (the sycamore—*ficus sur* or *ficus vasta*) which is a common *wodajja* venue, is considered sacred by Islamic cultures in many parts of Wollo and the northeast. The word "warka" (sometimes pronounced "werka") connotes, according to one recent Amharic dictionary (Desta Tecklewold) a place of ritual prayer. Springs and streams are also revered because they are thought to be the abode of the good spirits as noted above. Men and women often will not cross brooks or running water except on certain hours in the day for fear of disturbing the spirits that live there and drawing their ire. Water is also thought to be sacred because it has cleansing and healing properties.

The *wodajja*, in which both men and women participate, is held in the afternoon, and people gather and sit in a circle. The seating arrangement follows the rankings given to individuals by the community: the inner circle will consist of the local leadership, and on the outside the young and the less prominent sit. Coffee is prepared and drank, *chat (catha edulis)* is distributed until the time comes for the high point of the ritual, the Sacrifice. The officiate in this ceremony is the *abegar*, the traditional "political" leader in rural Moslem Wollo.

Unless the man of knowledge specifies what is to be sacrificed, the community will select its own animal. All sacrificial animals have to be male and of a certain colour or combination of colours.[1] Often white oxen, red, brown or white goats are preferred. If the community cannot afford an ox or a goat, it may sacrifice a white or brown cock or cocks. The meat is distributed to everyone present and consumed there. The *wodajja* may be repeated as many times as the community wishes. Some peasants in the Worebabo area had participated in four or five *wodajja*, others as many as seven in 1984 and 1985.

Extension agents in Ambassel reported that during both 1984 and 1985 a considerable number of animals were slaughtered and consumed in the rural communities. This they thought was an attempt by peasants to minimize their loss: it is better to eat one's animal than lose it through death. While it is difficult to estimate the number involved, a good proportion of the cattle slaughtered were for sacrificial purposes, and as part of the propitious ritual carried on in both religious communities.

Summing up, it should be clear from the foregoing that no effort is spared to determine the behaviour of the weather and to divine the probable course of natural events as these relate to the primary livelihood of the peasantry. If indigenous disaster forecasting proves inadequate, and it frequently does, it is not for lack of effort and concern, but rather because the parameters that are employed are traditional, the measuring rods are inadequate, and the underlying conceptions are at times prescientific. Be that as it may, all the peasantry's intellectual, social, religious and magico-mystical resources are deployed to anticipate and prepare for disaster. It should be stressed that disaster anticipation is a collective effort, and the practices employed are frequently group-oriented and fall within what we have called the cooperative ethic.

[1] A similar sacrificial ritual exists among the Sidama, but here cows may be slaughtered for the occasion (Brogger 1986: 137–141).

Crisis management

Let us begin with a few remarks about agricultural conditions, and the impact of the weather in Ambassel in 1983 and 1984. There are several developments in the area that should be noted at this point because they have a bearing on the course of events in the famine years of 1984–1985, and have to some extent affected the response of the peasantry to the crisis.

The Haiq area is probably representative of the *woyna-degga* zone (see Chapter 3 for the various zones) of Ambassel in agro-ecological terms. The rainfall data for the area for 1983 and 1984 (Annex 1) actually shows that the failure of the *belg* crops in the previous year was not so much due to drought as—and this is not unusual—to too much rain which simply smothered what was planted, and also destroyed farms. The *meher* rains, on the other hand, were irregular and inadequate, leading to crop failures on a large-scale again. The ensuing year, 1984, records the lowest rainfall in the area's twenty-two year history, with severe shortage of rains in both seasons. However—and this is the main point we are driving at—in both years *there was some rain*, and the awraja did not suffer absolute and complete drought.

Now, all things considered, too much rain is better than too little rain, even though both may result in crop failures, and conversely, some rain is preferable to no rain at all. The main reason has to do with livestock survivability which is enhanced if there is precipitation; the more, the better. Thus, while a large number of peasants suffered crop losses in both years, livestock losses, through deaths or forced sale, were reduced considerably as a result of the available moisture. The same holds true for several awrajas in Wollo.

The second observation to be made is this: precisely because of the availability of some precipitation in both years, as well as 1985, some peasants were able to gather a harvest, and thus some food was available in the awraja. The Red Cross study noted above (Vol. III, p. 26) shows that in Ambassel and neighbouring Qallu awrajas, where a crop survey was conducted, peasants suffered a 73 per cent loss in crop yield in 1984, but that 66 per cent of the farms in the area were actually cultivated. The study reveals that in the same year in Ambassel awraja some 25 per cent of the peasants in the survey had food harvest stored to see them through (most probably at reduced levels of consumption, although this is not stressed in the study) a period of time ranging from three to twelve months. Breaking down the figures into lower aggregates, about 11 per cent of the respondents in the awraja said they had enough harvest for three months, 10 per cent had a harvest to last them six months, and 3 per cent to last a year. Approximately 75 per cent of the peasants interviewed here had no harvest at all (ibid: 31).

Similarly in Qallu awraja, one of the worst hit awrajas of Wollo nearly 10 per cent of the respondents said they had a harvest in 1984 to feed their families for three to six months, and 90 per cent said they had no harvest at all. In the former group, 8.7 per cent had food for three months, and 1.7 per cent for six (ibid.).

This point must be juxtaposed with another important fact, namely that some areas of Wollo in the neighbourhood of or with traditional links with Ambassel, notably western Dessie Zuria, Wore-Illu, and Borena awrajas, suffered very limited crop losses in both 1984 and 1985. These areas are, traditionally, food surplus and food "exporting" areas, and peasants from here were active in the grain and livestock trade in the affected areas during the crisis.[1]

A third observation has to do with the timing of the crop failures, and of the ensuing famine in the province. We may speak of the "staggering" of the food crisis, by which is meant a condition in which some areas succumbed to famine earlier than others. In other words, some peasants had food in store even at the height of the famine while others were in the throes of hunger as early as 1983 or even earlier. The selective or staggered impact of the crisis was also evident within an awraja, or smaller territorial areas such as a woreda, or even a kebbelae (or PA).

Thus, a good proportion of the *degga* and higher *woyna-degga* areas of Ambassel escaped the terrible effects of the drought in 1984, while the lower ecologies of the awraja were battered by it as early as 1983. At times one local community would be in deep trouble while its next-door neighbour would be suffering only minor damages.

An extreme example of the selective nature of the crisis was what happened in a kebbelae in Ambassel woreda, told by the officials we interviewed in Wichalae, the capital of the woreda. About a third of the peasants in this peasant association, which was called PA No. 033, suffered heavy crop losses in both seasons in 1984, while the rest managed to gather a moderately good harvest in the main season of the

[1] For the famine population of the various awrajas of Wollo, see Annex 3. There was no famine in Borena and Wore-Illu awrajas in 1984.

same year. Interestingly enough those who suffered hardships were those who were located in the higher elevations of the kebbelae; the majority who lived in the lower elevations were saved from disaster not because they received more rain but because they had access to sources of water (several streams which failed to dry up during the year) which they could use for farming purposes. In the other cases noted above the difference was a function of differential precipitation.

The point we are emphasizing is that the famine of 1984/85 involved, at least in areas located in central and southern Wollo, not the absolute but rather the relative scarcity of food for humans and feed for animals. Additionally, the "staggering" or selective impact of the crisis played an important role in peasant survival strategies in the months before emergency relief became available to the needy.

Crisis survival brings sudden and dramatic changes in peasants' economic thinking for a brief period. In this phase of the crisis we see the peasant desperately trying to break out of the subsistence system, and plunge into the cash economy and the exchange system. The market now assumes greater importance than before as the peasant becomes aware that survival depends on acquiring cash to purchase food.

The value of an asset is now determined by its saleability, and the speed with which it is turned into cash. Barter is shunned as strongly as hard cash is sought. An asset will have high saleable value: (a) if it is easily transported and easily converted into cash; (b) if its value does not depreciate dramatically as a result of over-supply; (c) if the asset itself does not deteriorate. At a certain point in the crisis farm animals become poor investment as the market becomes flooded with livestock and the prices offered for them reach rock bottom, and as the animals themselves deteriorate and sicken due to the lack of feed and water. In the months of August to November 1984, the Friday livestock market in the town of Haiq was overflowing with farm animals: on the average 1,000 heads of cattle, and nearly 2,000 sheep and goats came to be sold every week (see Annex 2A). As a general rule, in crisis circumstances a community with a more developed market system, and a greater involvement in the exchange process, has a better chance of survival than a community where the opposite holds true.

The exchange process is not an exploitive mechanism in and of itself, indeed, it is the means by which the owner of one commodity meets the owner of another. Those who have argued that it is the capitalist exchange system that exacerbates famine often forget that it is the peasant himself who is in great need of the market, and who actively participates in it during a food crisis. Moreover, the experience of Ambassel reveals that by and large exchange during famine takes place: (a) among peasants themselves; (b) between peasants and rural traders; in a number of cases, rural traders are peasants who farm and engage in business alternatively; (c) between peasants and urban-based merchants. These are the same forces that play a key role in the exchange process during stress-free times.

However, the food deficit peasant is at a disadvantage in these circumstances (note that it is not all peasants but the *peasant short of food*) because what he has to offer is not in demand while what he wishes to acquire, i.e. food, is both relatively scarce and in high demand. On the other hand, the peasant with surplus food and the rural trader stand to gain in the exchange process. If the commodity traded is livestock it is the merchant and the rural livestock trader who benefit. In brief, in famine conditions the food deficit peasant has everything to lose and nothing to gain.

The saleable items frequently seen in the rural market during famine are a polyglot of all types of possessions. They include, apart from farm animals, agricultural tools, clothing, family furnishings and heirlooms, jewelry, riding and packing harnesses, hand weapons, craft products, and firewood. A practice which was fairly common in Ambassel was the selling of one's dwelling, usually done piece by piece. The information we have on this is not complete, but it may not be an exaggeration to say that between 15 and 20 per cent of peasant households in the hard hit areas of the awraja systematically stripped and sold their homesteads during the famine. Some of the building material was sold as firewood, the rest as material for new homes. The chief extension agent in Worebabo told us that a good number of people in Bistima, the woreda capital, built new homes, taking advantage of the situation of distressed peasants.

We shall deal with the system of livestock sales adopted by Ambassel peasants in Chapter 8.

The other major indicator of the departure from the subsistence ethic on the part of peasants during famine is the sharp increase in the demand for wage employment. This heightened awareness for wage labour triggers what we call temporary migration out of the rural areas. Where peasants migrate to, how early the process begins and who migrates and who does not depend on a host of factors of which the following are noteworthy: (a) The sources of employment and their location. The argument that peasants in distress automatically flock to the urban areas does not always hold true. Where the traditional sources of seasonal employment are located is where peasant job-seekers are attracted to, and these may be big plantations, large-scale rural construction schemes, contract farming enterprises, and the like. (b) The composition of the individual household, particularly its age structure. If a household has surplus labour power, i.e. if there are family members of working age, it is more likely that one or more persons from the family will migrate. Where this is not the case, migration may be postponed until the last moment, i.e. the moment of death and dispersal.

It is usually more difficult for the head of the family to migrate, and if he decides to do so there will be a number of problems he will have to deal with before he leaves. He will have to make sure that there is someone in the family who will farm his land should the weather change for the better in his absence; quite often an arrangement to this effect is made with a neighbour, friend or relative in the kebbelae or local community. He will also have to ascertain that his land will not be taken away in his absence. This is a constant threat for all peasants even in times free of crisis. He will also have to make arrangements that children and youngsters in the family have something to fall back on.

As the data further on shows, Ambassel peasants were reluctant to leave home and seek employment elsewhere. Apart from the difficulties noted above, the sporadic rains that fell in the areas in 1984 raised hopes among peasants that the weather would improve at any moment and make farming possible. Furthermore, migration was not an attractive proposition because the opportunities for off-farm occupation had deteriorated considerably since the revolution.

On the other hand young peasants, who have only recently established their homesteads are more likely to migrate as they are less burdened with family responsibilities. Family connections elsewhere and/or previous urban experience is also an important factor in migration. All things considered, peasants with relatives elsewhere or with extended urban or migratory experiences will find it easier to leave home in search of work than those without relatives in distant places or with little or no previous experience. A.P. Wood's argument (1976) that there is a connection between class status and migration is no longer valid as the land reform has brought about a radical levelling down, and has done away with gross inequalities in land possessions.

The decade leading up to the revolution in 1974 may be called the best decade for peasants in northeast Ethiopia in so far as off-farm seasonal employment is concerned. This is the decade which witnessed a high level of migratory movement by Wollo peasants both to engage in seasonal labour and to settle elsewhere. We met Wollo peasants in Wollaita-Soddo in Sidamo province, and in Manna woreda in Kaffa province while doing field-work for an earlier study. In Manna woreda one of the kebbelaes we selected for our interview was known by the name of Wollo *Seffer* (i.e. Wollo Settlement) because the locality used to attract a large number of temporary settlers from Wollo who came to work mainly as coffee pickers and loaders; the kebbelae still has a small number of Wolloyaes at the time of our field-work in 1980. Wollo peasants also travelled as far as Illubabor province both to settle and to work as labourers in the wild coffee plantations in the area (the best work on spontaneous migration from the highlands before the revolution is A.P. Wood 1977). Every year tens of thousands of peasants from Wollo would migrate to the large modern plantations and to the coffee growing areas in pursuit of economic gains. The income thus acquired was meant to serve as cushion against hard times.

Among the major sources of seasonal employment for Wollo peasantry were the two giant agricultural schemes on the eastern and northwestern lowlands of the country, namely the cotton plantations of the Awash valley, and the sesame seed farm enterprises in the Setit Humera region. Cossins has estimated that some 20,000 Wollo peasants were annually employed as seasonal labourers in the farms in the delta of the Awash river (1974: 59); there were also large numbers of them working in the Humera area, and in southwestern Ethiopia in the early 1970s (see also Cossins 1973). In an earlier work I estimated that the Awash Valley and the Humera enterprises employed over a quarter of a million permanent and temporary workers originating from outside the regions in the same period, and about one-third of these, i.e. some 70,000 to 80,000, were from Wollo (Dessalegn 1986 A: 81). In the Awash Valley in particular Wollo peasants were employed as "outgrowers" which involved raising crops on rented land to sell to the large plantations on a contractual basis.

The opportunities for off-farm employment for peasants in the northeast region, as in other regions in the country, are now bleak. The Setit Humera agricultural complex collapsed about five years ago and is today only of historical interest. The Awash Valley plantations, many of which were nationalized by the government after the revolution, are now a pale shadow of their former selves, and the seasonal work needs of those still functioning are met by means of state-sponsored commandeered labour. Peasants in many awrajas of Wollo are forced to work in plantations in the Awash Valley in the summer months without payment, and as a form of periodic "labour contributions". Needless to say, the peasants in Ambassel and Dessie Zuria awrajas we talked to about this were resentful of it. Finally, there are hardly any industries in the urban areas of Wollo, or of the adjoining provinces, to attract peasant job-seekers. In short, purposeful migration during times of stress is nowadays a subdued affair.

About 22 per cent of the respondents in our awraja said they had migrated out of their home areas in search of employment during the height of the crisis in 1984. Of these an insignificant number said they had travelled out of the province itself; two peasants said they had gone to Addis Ababa, and another two to the port of Assab in the docks there. Most of the migration was thus confined within the province, and a greater portion of the migrants had stayed in the rural areas, as shown in Table 22.

Table 22A. *Number of household heads who migrated for work (to urban & rural areas)*

Woreda	Urban	Rural	Total
Ambassel	7	5	12
Worebabo	5	10	15
Tehulederae	9	19	28
Total	21	34	55

Table 22B. *Number of household members who migrated for work*

Woreda	Urban	Rural	Total
Ambassel	4	5	9
Worebabo	13	23	36
Tehulederae	–	19	19
Total	17	47	64

Consider the two tables for a moment. Of the household heads who left their homes to look for work, 62 per cent stayed within the rural areas; the comparable figure for household members is 73 per cent. This is interesting because the latter group being younger and carrying less family responsibilities would be expected to venture further away from home. The implications of the data from the two tables as far as migration as an early-warning indicator is concerned should also be evident. A greater part of peasant migration in times of crisis takes place within the rural areas where it remains undetected until the crisis reaches a point where migration in search of work is replaced by mass displacement.

Of the 55 household heads who left home to look for work 31, or 56 per cent found some form of employment, agricultural work being the most important. Daily labour in the urban areas, handicraft work, selling firewood, doing odd jobs ("shiqqella" in the local vocabulary), petty peddling, and domestic work for women were the other types of work found by job-seekers in order of importance.

Let us return to the market system and examine the behaviour of prices in the rural areas for the years 1984 and 1985. As was noted earlier the peasant deficit in food suffers from unfavourable "terms of trade" as what he has to offer for sale is in low demand and what he needs in exchange, namely food, is priced very high. Annex 2 and 3 show the movement of prices of cattle in Haiq, and of grain in Bistima (Worebabo woreda) market during a critical period in the famine. Unfortunately the two tables are not fully comparable as the dates on which the information was collected varies considerably from one to the other. Additionally the Bistima grain data, collected by the Save the Children office in the area was collected after emergency food aid had begun to arrive in the area.

Despite that, however, the price of grain, including that of wheat, which was the main cereal food distributed in the area by relief agents at the time, continued to climb for many months.

The price of the chief local crop, sorghum, remains high up to October 1985. Indeed, at one time it fetched 204 Birr per quintal; in a normal season sorghum would be priced between 40 to 60 Birr per quintal in the local free market. Conversely, the price of an ox in the first two quarters of 1985 was between 125 and 150 Birr, and was just recovering from a sharp fall in 1984. In the autumn months of that year the market value of an ox fell as low as 50 Birr, that of a cow fell to 35

Birr, and calves were quoted at 15 Birr per head. In 1983, an ox was priced in the Haiq market 150 and 250 Birr; in a good season a farm ox in good condition would not be available in the larger Ambassel cattle markets below 300 Birr.

Let us return to the food price table, Annex 3. The price of pulse crops remained as high as, and at times higher than, the price of cereals during the period covered by the data. Pulses are as important to peasant diet as cereals, and this fact has often been overlooked in the relief effort which concentrates on providing cereal food. On the other hand, the volume of cattle in the awraja capital market (Annex 2) each month rises sharply beginning in June 1984, and reaches a high point in October–November of the same year. Interestingly enough also, the market is still flooded with cattle (as well as with sheep and goats though not shown here) for much of 1985 despite the fact that relief assistance was being distributed at a high rate in the area at the time, and some rains had fallen in the *belg* and *meher* months.

Several reasons may be offered to explain this. First, the relief effort was limited, and many peasants had not yet fully benefitted from it. Second, peasants were burdened by extraneous obligations, such as taxes, state levies, loans, etc., for which they had to sell livestock to raise the necessary cash. Thirdly, peasants were still wary of the return of the weather to normalcy, and thought it wiser to hold cash rather than "perishable" commodities like livestock. They may also have had fears that the main autumn harvest would be below normal or even poor.

Where did the food sold in Ambassel rural markets come from, who were the suppliers, and what kind of food was available in 1984 and early 1985? To answer the last question first, there is reason to believe that a majority of the food crops commonly consumed by the peasantry were available, though on a greatly reduced scale, in the rural markets of the awraja in this period. However, in many parts of the awraja sorghum and maize became hard to get in the first quarter of 1985, and their prices shot up sharply in this period as a consequence. In the last quarter of 1984, and the first quarter of the following year, barley, peas and lentils (the last two important items in peasant diet) virtually disappeared from the market. However, some substitute crops, for example *guaya (lathyrus sativus L.)*, a lentil substitute, was available in some markets in the awraja (NEERNDRC May 1985 AA: 58).

As was noted earlier the food sold in the Ambassel markets came from two main sources: (a) from the food surplus areas of Borena, Wore-Illu, and western Dessie Zuria awrajas; the latter two were more important suppliers of food to our awraja than Borena whose peasants and traders sold mostly in the Qallu awraja markets; (b) from peasants in Ambassel awraja itself, i.e. from those who had not been seriously affected by the drought, and who thought to benefit from the high prices offered for food in the local market, and the low price for livestock some of which they also bought. The suppliers active in the crisis period were peasants themselves, and rural traders, although in some areas urban-based grain merchants also played a part. On occasions food deficit peasants themselves travelled to the food surplus area to buy food at prices lower than those offered in their own localities.

It may be appropriate here to return to the subject of the family and see how the members, individual and as a unit, coped with the crisis during this phase. Before we do that, however, a few words about the treatment of the subject in the literature. We have already remarked on, albeit briefly, the views of those writers who have dealt with the victimization of women during famine. Other writers have gone a step further and have tried to show systematic discrimination against women as an essential part of agrarian social values in Third World countries. The evidence for this comes again from the Indian sub-continent, and the substance of the arguments may be summarized as follows: (a) Women, specially female children are deliberately neglected even in stress-free times in male-dominated Indian and Bangladeshi societies. This neglect and discrimination manifests itself in less food allocations to them as a matter of course; the result is higher levels of malnutrition, higher rates of mortality among girls than boys. More is invested in boys because they are believed to have greater economic potential, and higher income-earning capabilities than girls; families also believe that boys are more likely and are in a better position to support them in old age than girls. (b) In times of food crisis this bias against female family members is accentuated leading to higher rates of mortality among them than among the menfolk (Chen *et al.* 1981, Rosenzweig and Shultz 1982, Kynch and Sen 1983, Torry 1986).

It is of course a fact that peasants (often male peasants) in rural Ethiopia prefer male rather than female children, mainly for economic reasons. This is not as unjustified or as unfair as it may appear or is made out to be. A family in old age needs someone to help it farm the land, and without the labour of male relations the family may become dependent on others or may be reduced to penury. The preference for males rather than females used to be more accentuated in communities in this country where women did not inherit land than in those which did, but the land reform has done away with that difference now. However, conscious sex-bias in matters related to food allocations is not widely practiced in rural Ethiopia.

As was noted earlier the division of labour between male and female is distinct in the area of housework, and starts from childhood. The female child in highland Ethiopia begins to participate in domestic work as soon as she is old enough to do odd jobs in the home, to fetch water, and handle food. The processing, preparation and dispensing of food is an arduous task, and a girl begins to learn the skills from an early age. Moreover, women serve men during meals, and on any occasion where food and drinks are consumed. Older sisters for instance will serve younger brothers. It is often the case that during mealtimes girls and women eat less than boys and men, and some have taken this as evidence of sex-based differential allocation of food, which is not the case.

The food habits of rural women are somewhat different from their menfolk. Female etiquette demands that women take less food and eat more slowly in the presence of the men and of guests. On occasions, when important guests are around, women may only serve the food and may not themselves participate in the meals. However, women take food out of the presence of the male members of the family, and while preparing and dispensing the daily meals. The men have control over the production of food and some have mistakenly taken this to mean that they thus have absolute control over all food that is consumed in the household. The control of the processing and dispensation of food is the sole responsibility of women, and this gives them far more power in the family than is usually acknowledged.

In times of food crisis peasants believe the family has better chances of survival if it remains as a unit. The experience of Wollo suggests that famine does not lead to high rates of divorce because the services of each adult member become more important as the crisis escalates. That famine does not in itself lead to unusual or abnormal matrimonial behaviour is also suggested by M. Vaughan (1985) in the case of Nyasaland, although she does show that in the 1949 famine a good number of women lost their husbands because they failed to return home after travelling to distant parts of the country to look for food. Quite likely, however, the same thing would have happened in non-famine circumstances if a major event or crisis like war or unemployment had triggered large-scale out-migration as in this case. It would be interesting to know for instance, what percentage of the migrant workers who flock to South Africa in search of employment from the neighbouring countries return to their original wives.

It is true, however, that famine does exacerbate pre-existing family conflicts and uneasy filial relations, which at times may lead to family break-ups. We talked to several relief agents who had worked in some of the big shelter camps in Wollo during 1985, when the camps were active and the inmate population was predominantly female. Unfortunately the information they provided us with was impressionistic and lacking in detail, and we cannot use it for comparative purposes or to indicate a scale of magnitude. Some of the relief agents who were active in north Wollo remember being told by women of having abandoned their husbands and gone off with some of the children in search of assistance. A good number of these women, according to these agents, had no intention of going back to their husbands. One relief agent who had worked in the Bati area was told by several women in the camp that they did not know where their husbands were at the moment and were not too eager to find out; a few of the women, the agent said, had planned to go and live with their relatives who were located in a different area. Another relief agent whose work involves reuniting abandoned children with their parents informed us that the number of children who had abandoned their families and gone to the shelters during the famine was not insignificant. In all these cases, the underlying cause of the break-up was not so much the famine or the human dislocations brought on by it as long-standing unhappy relationships which had been smouldering for a good many years, and which break down when food crisis puts severe strains on men, women and children.

There is however a specific division of labour in the family occasioned by famine. Men are expected to look for wage employment, to do odd jobs to earn cash, and to travel to other areas to request assistance. The selective selling of cattle and the monitoring of the market is also mainly the work of men. Both men and women use their knowledge of natural resources to collect famine foods and edible wild plants. However, women are entrusted with the children, particularly the younger ones, and when the situation deteriorates they may leave home to engage in begging, often taking the children with them. The reason women are entrusted with the younger children (the older ones, those above six or seven may stay with the father) has nothing to do with female discrimination but is a result of the traditional division of labour in rural society. Men traditionally do not participate in the physical aspects of raising children, and do not have the skills of handling, feeding and generally caring for them. It is the task of women to do all this, and it is they who develop the necessary skills. Moreover, while begging in times of hunger is done by both sexes, women have a better advantage as alms givers are more sympathetic and responsive to the pleas of women with small children than to men.

The crisis division of labour very often drives the men and the women in different directions as each goes away to look for help, and they may stay away from home for several months. The general practice however is that as soon as enough assistance is found or the conditions in the home area change for the better both husband and wife (and the children) return home to resume their lives. A woman respondent in Tehulederae woreda told my field assistant that both she and her husband returned home on the same day after having travelled in different directions for several months in search of help. Several peasants in the same woreda informed us that they were reunited with their wives, who had gone off with the children to seek assistance, in the shelter camps several weeks after the latter had left home.

A family reduced to penury, and with death hanging over it will allocate what little food it has to children and mothers first, and the adults will frequently forego a meal or two in preference to the more vulnerable members of the family. Adults, particularly men, will engage in frequent religious-inspired fasts and this helps to stretch the stock of food available for the family a little further. Youngsters will scour the countryside and the woods for wild berries, roots, and other edible plants which again helps the weaker members of the family to get a larger share of the food at home. Peasant cultural values put emphasis on large households and the continuity of the family progeny, therefore children and mothers with infants are not victimized.

There is no evidence that in the 1984/85 famine traditional moral precepts were shattered in Wollo or that peasants indulged in abnormal or inhuman behaviour, as some have alleged. The experience of the people of northeast Ethiopia in the last two great famines reveals that peasants prefer to die in dignity rather than break deeply held moral values, or behave in strange and unaccepted ways. In an interview with Hancock, M. Amin, the man who made the famous BBC film about the

Ethiopian famine, reports seeing 150 starving people being fed in a compound in Meqellae (Tigrai province) while around them were over ten thousand equally starved peasants calmly watching and doing nothing. Amin states that if he were in the position of the latter group who were virtually being condemned to death, "I would not have just stood there and watched others being given the chance of life. I think I would have done anything, rioted, killed, to get the food I need" (Hancock 1985: 9).

Let us bring this sub-chapter to a close with some brief remarks on the role traditional associations and rural mass organizations played during the food crisis. We may begin with religion and religious institutions. Religion stands as the source of spiritual solace and of hope in rural society, and more and more people turn to it in times of crisis; this was true in the case of the peasantry in Ambassel in 1984 and 1985. Apart from providing spiritual comfort, however, neither the Christian nor the Moslem religious institutions played any active role in mitigating the crisis or providing assistance. Religious leaders did of course support the rural population in their prayers and other religious observances, but outside this, organized religion was just as helpless in the face of death and starvation as were other traditional and modern social organizations.

The traditional mutual-benefit associations in Wollo are *mahber* and *idir*. The first is an association which is observed periodically, and in which each member takes turns to prepare a modest feast and to host the gathering which often takes place monthly. *Mahbers* may be organized on sex lines, there may be a *mahber* exclusively for the men and another for the women in a community. These associations are often burdensome because considerable energy and resources are spent in the preparation of the feasts. The second is an association of mutual support in times of family deaths and loss of relatives. It is simply organized, does not often require large outlays from members, and plays a very important role in rural society.

According to peasant informants, the *mahber* was quickly phased out in many communities as more and more peasants found it difficult to meet the obligations. Some peasants said that the *mahbers* were not abandoned or written-off for good, but merely put in cold storage, as it were, until the conditions improved and peasants could look forward to a normal harvest. One peasant in western Tehulederae woreda informed us that he hoped to reactivate the association in his community as soon as the current harvest was gathered in. *Mahbers* often have religious significance and it is this aspect which drives peasants to be involved in them.

The *idir*, however, continued to be observed in many communities even at the height of the famine until the moment the communities themselves were abandoned. It was the *idir* which helped bury the dead, supported families in their bereavement, and by mobilizing the cooperative efforts of the community took care of most of the ritualistic practices associated with death and mourning. Without the association many peasant communities would have been helpless in the face of mounting death rates, and the risks of an epidemic would have been great as more and more of the dead would have been left unburied.

According to some peasants and peasant association leaders the first institutional casualty in the face of the escalating famine was the peasant association. In a good number of communities the PA simply vanished, and as one peasant in Wichalae put it, people did not know where it disappeared to ("yet i'ndeggeba alawoqn'm). In some cases a number of PA officials, who were themselves suffering from the famine, abandoned their kebbelaes and went looking for employment elsewhere. As the crisis deepened those PA officers who had not migrated simply packed up and left with the rest of the community to the urban areas and the shelter camps. According to some of the PA leaders there was nothing for them or the organization to do, and the problem facing their communities was so large that they thought only the government could handle it.

Of the eight PA officials we talked to on the subject, three had left for Addis Ababa and Assab to look for work (each had a relative there), two had gone to Tendaho to work on the cotton plantation, and the rest managed to survive with the help of emergency relief. The peasants in the kebbelaes did not of course need the PAs, nor did they miss them when they disappeared. While the PA could have played an important role in mobilizing peasants to withstand the crisis better, its fragile structure and its subordination to the local state bureaucracies did not allow it to rise to the occasion. The PA, on the other hand, resurfaces at the time when emergency relief is set up in the rural communities, and in the period of post-famine recovery which will be discussed further on.

Exhaustion and dispersal

This is the final stage in the progress of famine when whole communities, involving tens of thousands of people, flee their home areas with whatever possessions they have (which is often next to nothing), and descend on the towns. At this point in the tragic drama, the death toll has considerably exceeded normal expectations, peasants have thoroughly exhausted all useful resources in their communities, and there is nothing except death to expect or wait for.

The journey from the rural areas to the road-side towns—these are the initial recipients of the refugees—may be short for some, but for the majority who live in the interior it may take several days. Travelling is slow as the party contains the old, the sick, and the infirm some of whom have to be assisted and some carried on people's backs. A large number of deaths occur in the process, and on many occasions the dead are given a quick and simple burial, and the traditional mourning ceremony is dispensed with. So much crying, weeping and anguished mourning has been done already in connection with the numerous deaths that have occurred in the weeks and months before the journey that men and women are drained of all emotions, and burial takes place in silence and with many people wearing wooden faces.

Crisis dispersal is a collective act. It is true that prior to this stage individual peasants or households may have decided to migrate, but mass migration, the kind that flooded the small towns of Korem, Alamata, Qobo, and Bati in 1983 and 1984 is often a conscious decision of the *got*, the kebbelaes or, at times, the local community. In a sense this should also be considered a cooperative endeavour. Needless to say, men and women help and morally support each other on the journey, and in the makeshift camps in the towns. Moreover, the wisdom of the act should not be lost sight of. Peasants understand the impact of numbers: individual destitutes may arouse the sympathy of individual city folk but are invisible to the local power holders, while a mass of starving peasants numbering in the tens of thousands is far more likely to attract the attention of the authorities.

It is ironic that peasants whose life-long occupation it is to feed everybody else should go to others to ask for food and sustenance. It is a strange reversal of roles indeed: those who produce food requesting food from those who do not! But a closer examination reveals that the problematic is more complex and more nuanced. The mass migration of peasants to the urban areas carries also a variety of meanings and messages. To begin with, it is a form of collective articulation of the demand for the consecration of the right to life, a right which requires the satisfaction of the most elementary needs of individuals and communities, namely food. Secondly, it is a form of silent protest. The migration of thousands of peasants from the vast recesses of the countryside where they are "invisible" to the roadside towns where they are highly visible contains a clear but unexpressed message: state and society have no political or moral right to write us off, or to ignore our plight! This form of collective articulation of grievances has meaning and force only when it is conveyed dramatically and in great numbers. Finally, mass migration is a form of escape from the spectre of death, and the taint and evil consequences of what may be described as spiritual pollution.

Many peasant belief-systems in this country contain elements about evil spirits and their infestation of settlements and other living space in certain circumstances. Nomadic pastoralists, for instance, will abandon a settlement and move elsewhere if they suspect that the location is infested with malevolent spirits. Shifting cultivators will do likewise if their religious leader reports that the existing settlement has attracted spirits harmful to the community. Spirit pollution is greatly feared among both sedentary and transient peasantry. Among the Wollo peasantry the community is believed to have become polluted or spiritually unclean when it suffers an unacceptably high number of deaths during a natural or social crisis. In this circumstance, the community is thought to have become the home ground of Death and Evil, and it is therefore considered necessary to flee from the locality to save oneself from being possessed by evil and from suffering an unclean death.

Why do peasants wait until the last moment to leave their communities and to migrate to the cities? Given our present knowledge of peasant psychology and moral behaviour, the question will have to be kept in limbo. The answers provided there are far from complete, nor do they fully explain the interplay of economic and social forces that impinge on rural producers and affect the timing of their exodus. It must be stressed, however, that it is under extreme conditions—and then only with great reluctance—that peasants will abandon their communities and become refugees or dependents elsewhere. The decision to leave one's place of residence and move to other areas, which, near or far, are often considered by peasants to be hostile or to provide very little welcome, is one of the most difficult decisions a peasant has to make in his lifetime. Add to this the fact that the insecurity of individual holdings which is a legacy of the land reform of 1975 tends to tie down peasants to their communities, it will not be difficult to see that family migration and dislocation can only be an act of desperation and finality. This explains in part the fact that in a number of relief shelters in Wollo there were a far higher proportion of female inmates than male, particularly in the period immediately after the initial phase of operation of the camps. Men peasants will on occasions send their family or womenfolk to seek help in the relief camps and stay behind to keep an eye on the farm; conversely, the menfolk may decide to leave camp earlier, leaving their families behind so that they can do some work on their plots, or plan for the coming farm season. Dependency is a difficult thing to accept for Wollo peasantry.

The reasons why peasants postpone the decision to pack up and leave their localities until the last and fatal moment are mainly two-fold. In the first place, part of the blame for the "bad" timing must go to the faulty "early warning" or "disaster forecasting" system of the peasantry. No system of forecasting, not even that which employs the most advanced technology, is fully accurate, and the system of forecasting adopted by the peasantry suffers from far too many weaknesses. Moreover, the belief that all disaster is a manifestation of God's will, (which is how peasants explain famine and ecological crisis) contains a kind of redemptive principle subsumed in it, namely that normalcy will return once God's purpose (which is often considered a punitive one) has been achieved. When and how a crisis will come to an end is of course not known to the peasantry, but the Supernatural explanation of disaster tends to arouse peasant expectations for an early end to disaster, and for a quick return to normal conditions. In 1984 and 1985, peasant hopes for an early end to the drought were kept alive by the intermittent rains and sporadic showers that fell both in the *belg* and *meher* seasons. These rains were of course hardly sufficient for cropping purposes but they kindled peasant spirits, and helped keep livestock alive in many parts of Wollo.

The second important factor which had a bearing on the timing of peasant migration has to do with what may be called peasant self-esteem. Among the peasantry in Ambassel, self-maintenance and family self-sufficiency are higly respected social values. The same is true among peasants in many parts of Wollo. A respectable peasant in our awraja is one who maintains his family through his own efforts, who is not dependent on others, and who is hard-working and frugal. Self-reliance in this sense does not of course invalidate the system of co-operation which has been discussed in this study earlier. The co-operative ethic which informs all aspects of peasant life, is grounded on reciprocal support, and on transactive functions of mutual benefit. The dependent peasant, the peasant who relies on others for his livelihood will sooner or later be excluded from the co-operative framework since he will be unable to contribute to the flow of services and benefits.

How important peasant self-esteem is and what role it may play in peasant decisions in crisis is indicated in Table 23. One of the questions we had in our questionnaire was: What will you do if serious drought and famine occur again? The answers we got clearly reflect the importance of the social values of independence, self-reliance and hard work. The reader should be reminded that in two of the three woredas of our awraja the questionnaires were filled out at or near the towns or centres where relief supplies were handed out, and in all three woredas the interviews took place at the time when peasants were collecting their monthly emergency rations.

Only 15 per cent of all peasants interviewed said they will seek emergency relief if famine strikes again, whereas 30 per cent said they will work hard, save and make better use of their resources in such an eventuality. If we include those who replied that they will seek paid employment, the figure for what may be termed the "activist group" will go up to 42 per cent. Those we may call "defeatists", i.e. those who were embittered enough to consider giving up everything, including their lives if another famine occurred made up only 5 per cent of the total. Interestingly enough less than 2 per cent of the peasants said they would consider going to resettlement in case of another drought. A more detailed breakdown of the answers is given in Table 23.

Table 23. *What peasants will do if another famine occurs (% of sample peasants, Ambassel awraja)*

1	2	3	4	5	6	7	8	Total
15	12	30	1.6	5	5	14	18	100.6

Note: 1. Seek emergency relief. 2. Seek paid employment. 3. Activist: diligence, frugality and better use of resources. 4. Go to resettlement. 5. Drought will not occur again. 6. Defeatists: give up farming altogether, give up and die. 7. Fatalists: God or Allah knows, we will pray. 8. Other.

The table reveals, albeit indirectly, the general attitude of Ambassel peasants with regard to the question of migration and dispersal. Of the more than seven choices selected by our respondents (we shall leave out column 8, Other) two and a half, i.e. columns 2, 4, and half of 6, involved leaving one's local community. Even if we include column 1 in this categoy—being a recipient of relief does not necessarily lead to migration—only about 32 per cent of peasants chose alternative courses of action which would lead to abandoning their communities, whereas more than half indicated choices (some of which were actually choices of "inaction") which would not involve migration as a consequence.

To many peasants the ability to struggle against adversity with one's own means, and to hold out as long as possible is a source of pride and serves to enhance a person's standing in his community. The net result, however, is that the decision to abandon one's *got* or kebbelae is put off until the critical moment when any other course of action will only lead to death.

Hungry peasants have always flocked to the towns where they put up makeshift shelters and silently await their fate. The involvement of relief authorities, local and foreign, and the provision of emergency services to the victims of famine is a new experience in this country dating back to 1974 when for the first time "organized famine relief" played a critical role in helping peasants in the northeast recover from the worst disaster they had ever experienced. However, the evidence available indicates that this relief experience did not play a significant part in the decision of the Wollo peasantry to migrate to the roadside towns in 1984 or earlier. This is what hungry peasants in Wollo have done whenever disaster has struck, and this is what they did in the early phase of this last famine. As was noted earlier, the urban areas are the seat of authority, and authority in most peasants' minds is believed to possess unlimited resources, and it is this fact, more than anything else, that makes the towns a point of attraction to the destitute and the starving.

At the initial period, therefore, the relief camps go to where the people are and not vice versa. The shelter camps in Korem, Qobo, Bati, etc. were built by local and foreign NGOs to cater to the large refugee population which was camped in these places, having migrated there from the surrounding countryside. Once the relief process is under way, however, the camps become points of attraction to the peasantry, and are sought out by those in need of food aid and medical services.

Can it be said that the shelters, and emergency assistance in general, has had the effect of breaking down peasant resilience to adversity, or of encouraging the rural population to become dependent on external interventions? Is it possible to suggest that peasants did then, and will continue in the future, to look outwards rather than inwards, and to pay less attention to their own indigenous techniques of famine survival? These are important questions that need to be examined in depth, for the answers that may be obtained will have important implications for rural policy and future emergency operations.

It is unfortunate that not enough time was available to us to explore the problem of the effect of relief activities on peasant attitudes and expectations. It is our estimation, however, that the impact of relief on indigenous survival strategies is less damaging than has been suggested by extension workers in Wollo, and should another food crisis occur it is highly unlikely that the peasantry would promptly turn its back on its traditional coping practices and wait for outside benefactors to rescue it. To many peasants in Ambassel, as well as those in many parts of Wollo, relief appeared at the moment when the resources of their own communities had been exhausted, and when mass migration had already begun in earnest. External intervention arrived, in other words, after the traditional strategies had proved to be inadequate; it did not supplement or come to the aid of indigenous survival operations. The choice, thus, for a majority of Wollo peasantry was not whether to continue to cope by their own means or seek emergency relief, but rather to go in search of assistance as there was nothing else to be done.

8. Survival Strategies

In Ambassel awraja, as in the northeast as a whole, every peasant's life history is punctuated with suffering and hardships arising out of large-scale endemic hunger, and every household has invariably lost a family member, a close relative, or a valuable friend in one of the numerous famines that have struck the region in the last four decades. The majority of the peasants in our questionnaire, for example, had experienced four serious food shortages in their lifetime, while the rest had suffered through at least two major famines. No family, and no individual has escaped the agonies of hunger, or the consequences of it, and though the evidence may not be readily apparent in all cases, there is no doubt that everyone in the region has been seared by these tragic experiences.

To a peasantry living in acute destitution and imminent danger, survival considerations are always paramount, and every peasant learns the techniques of survival as part of his/her everyday experience. These techniques may be crude or ingenious depending on the frequency of disasters experienced by the peasantry, the perceptions of the people and the stock of accumulated knowledge having to do with production and survival, the resources (natural and social) of the community, and the social relations and communal values existing at a given time. Tawney was greatly impressed by the survival instincts of the Chinese peasantry in the face of countless disasters through the centuries. It is, he says, "difficult to resist the conclusion that a large proportion of the Chinese peasants are constantly on the brink of actual destitution. They are, so to say, a propertied proletariat, which is saved—when it is saved—partly by its own admirable ingenuity and fortitude, partly by the communalism of the Chinese family, partly by reducing its consumption of necessaries and thus using up its physical capital" (:72).

The same thing could be said of the peasantry of northeast Ethiopia, although the specific measures employed in coping with disaster may vary in detail, if not in substance, from those adopted by the Chinese peasantry. Ingenuity, strength of character, effective use of natural resources, and communalism: these are what have enabled the rural population of the northeast to live through large-scale disasters, whether these may be the product of natural causes, of political calamity or economic misfortune, or a combination of both. It is tragic but true that the two major elements that are essential to the well-being of the peasant mode of production, namely political and social stability on the one hand, and environmental constancy on the other, have often eluded the peasantry of the region. Indeed, it has frequently been the case that natural calamity has given rise to social disorder which in turn has fuelled privation and hunger in what seems to be a never-ending cycle of human misfortune on a grand scale.

Indigenous disaster survival here involves the adoption of emergency-induced resource management measures, the effective use of natural resources, divestment of savings and disposal of assets, and greater and more efficient use of the market system. Cutler and Stephenson (1984: 5) have drawn what they call a "model of peasant behaviour under famine conditions", which they say was applicable to northern Ethiopia during the famine years of 1983–84. "Peasant farmers initially react to crop failures", it is argued, "by (i) migrating in search of work in crop-surplus areas outside the famine zone (often in large numbers and usually far afield); (ii) by going into debt and (iii) by selling off less important assets such as small stock (especially sheep and goats)". This, it is pointed out, has the effect of driving up grain prices and pushing down livestock prices. Wages tend to be depressed, and soon whole household migration out of the famine zone takes place (see also Holt and Cutler: 9–10).

The discussion presented in the previous chapter is sufficient to show that this "model of peasant behaviour" did not actually hold true for Wollo in the period under discussion. As was argued earlier most of the peasants we interviewed in Ambassel awraja were not involved in migration. Of those peasants who migrated 20 per cent travelled to distant places in search of employment, while 80 per cent stayed *within* the famine zone. A good number of these migrants stayed within the awraja itself, while a large majority travelled to the neighbouring awrajas of Qallu and Aussa, two of the most severely affected areas of Wollo province.

Secondly, there is reason to believe that rural money-lending dries up in times of serious food shortages, as the risks of lending are very high—borrowers might die of starvation, or permanently migrate elsewhere—and the credit worthiness of peasant borrowers falls sharply owing to their distressful conditions. On the other hand, rural lenders re-emerge, and lending activities increase at the end of the crisis and in the period of post-famine recovery, as we shall see in the next chapter.

The elements of indigenous famine survival may be grouped into four sequential series of activities, namely: austerity and reduced consumption; temporary migration (in our case this did not take place on a large-scale); divestment; and, crisis migration—i.e. mass migration, mass death, and wide-scale dislocation of communities. Assuming that the immediate cause of the food crisis was poor harvest in the *meher* (autumn) season, these four sequential stages may be periodized as follows: (i) austerity and reduced consumption: January to April; (ii) temporary migration: mostly in the spring; (iii) divestment: May to August; (iv) crisis migration: August to October. In this chapter we shall look closely at the first and third series of activities since we have dealt at length with the other two in the previous chapter.

Austerity and reduced consumption

This is where a peasant household's resource management system is put to its severest test as the needs of the family have to be carefully balanced with the available food resources which have to be used frugally to make them last as long as possible. Resource management here also involves sharply altering the mix of food items normally consumed in the family as well as reducing their variety and quality. The division of labour in the family remains distinct at this stage: while women are entrusted with the management of all immediately consumable food resources, men are responsible for managing other assets of the family, and for entering into reciprocal arrangements with other families for purposes of mutual support and exchange.

The most important elements of food management in rural Ethiopia are processing and storage of staple crops, spices, dairy products, etc., on the one hand, and preparation and allocation of meals on the other. However, before, or at the same time as these arduous tasks are carried out, women will try to augment the staple crops the family will rely on through the crisis by disposing of other food items in exchange. This kind of exchange may take place among neighbouring families by means of barter or may be done through the market.

As the crisis deepens and the household food stock dwindles, food management in general, and food allocation in particular, becomes an increasingly difficult burden on women. The menfolk now become dependent on the womenfolk as the latter become solely responsible for what food is prepared and how it is allocated to each member of the family. The women's task is made somewhat easier however by community moral values, which though perhaps not fully observed, continue to serve as a point of reference and guide to action in these circumstances. These values may not, and do not as a rule enable an individual to answer all difficult questions and determine all critical choices, on the contrary, their importance lies in the fact that they prescribe a minimal standard of behaviour, and indicate what courses of action are within these bounds, and what are unacceptable to the collective conscience.

Thus in both Christian and Moslem communities in Wollo, priority or special attention is normally given to pregnant women, children, women with infants, and the elderly, and this social "bias" continues to hold even in distress conditions. Such members of the family, or at least the first two, may thus receive relatively more or better food than the rest. It is not, on the other hand, uncommon for the elderly members of the family to refuse food, and insist that their share be given to children or the most needy.

Another custom that makes the task of women easier in allocating food is the practice, common in both Moslem and Christian communities, of abstaining from food periodically for religious or magico-religious purposes. As noted in earlier chapters, the adults in the family will engage in frequent fasts having to do with religious or proto-religious ceremonies and rituals conducted in connection with the famine. Needless to say, all customary group as well as religious, or religions inspired observances which involve the provision of food and drinks are abandoned temporarily, so too traditional practices of gift giving and exchange.

We were unable to collect information having to do with the effect of such austerity regimens on rural marriages. It seems likely that at this stage in the progress of famine the ritualist aspect rather than the volume of marriages will change. That is to say, there may not be a dramatic decrease in marriages since many peasant families will be eager to reduce the number of mouths they have to feed, and since it is to the advantage of the families of both bride and bride-groom to have their children establish their own homesteads. What will probably be different now is that weddings will become a more humble affair as families fail to afford large festivities or to exchange costly dowries or bride prices. More research on this is of course necessary to see if the Ethiopian experience approximates or diverges from, for example, the experience of the Indian sub-continent where postponing marriages comes as one of the earliest decisions of communities under distress (Caldwell, *et.al.* 1986).

The composition of the family will change during this phase of crisis survival. It is not uncommon for older families, particularly those which are without an able-bodied work force, to abandon their homesteads and move in with their married children. A good number of our respondents in Ambassel informed us that one or both of their parents had transferred their residence to theirs and had to be cared for during most of the famine. Single elderly parents (i.e. those who are separated or whose spouses are deceased) are more likely to move in with their children than parents who are not widowed or separated. There are now more mouths to feed than previously, and more pressure is put on the family. Initially, this practice may have some benefits for the combined family since it will enlarge the disposable assets available to the new household. The older parent's or parents' possessions, including the homestead (but not of course the land) will gradually be marketed and the proceeds used to feed the family. On rare occasions, the combined resources of both parents and children may be sufficient to see the family through a crisis.

As was observed earlier, one of the tasks of the (male) household head is to make arrangements with other families to acquire assistance or to set up mutually beneficial deals. Assistance often involves borrowing grain or other food stuff, while deals may take the form of barter-exchange, mortgaging assets in exchange for food, setting up credit arrangements, and the like.

In the most typical case, a peasant may travel to a neighbouring or distant community, which is untouched or only lightly affected by drought and famine to borrow food from a relative, a friend, or someone with whom such an arrangement can be made. If the food owner is unwilling to give his food on credit, the peasant may offer a variety of deal options:

(a) Offer to barter the food with an asset like a goat, sheep, hand weapon, etc.; the peasant will suggest this deal if it is better than selling the asset in the market.

(b) Pawn an asset, especially a farm animal (usually a female small stock) in return for the food. There are advantages to both parties here: for the food owner, he has an asset as collateral and will be entitled to the offspring of the animal if it calves in his custody; to the beneficiary, he gets food, and his animal will be able to survive the crisis.

Several of our peasant informants said they travelled as far as Borena awraja in search of friends and relatives with whom they expected to make deals or to obtain food. A good number of our respondents who were engaged in this practice did not however have to travel such great distances to seek assistance. Because of the selective nature of the drought in the famine years, there were pockets of communities in Ambassel (and also other areas) which had managed to escape the disaster, and peasants here were not unwilling to enter into deals with needy ones to obtain an advantage or a benefit. The areas in Ambassel awraja which were not seriously hit by the drought and which attracted peasants in distress were western and northern Ambassel woreda, and the *degga* and higher *woyna-degga* areas of Worebabo woreda.

These exchanges or deals are inter-family affairs and not open and routine business transactions. A stranger cannot walk to a "prosperous" peasant homestead and ask for food on credit or on barter. The exchanges are made between or among relatives or friends, or through third parties which have some connection with the food owner, recipient or both. This is of course a form of inter-family, and inter-community cooperation. In material terms greater advantages will accrue to the "prosperous" peasant, while in human terms (i.e. in terms of lives saved) the needy peasant stands to gain. Most food loans are payable with "interest", i.e. an agreed upon quantity of food is returned to the lender over and above the original amount at the time of payment; what the "interest" will be depends on the specific arrangements made between the parties involved, and the time it takes to pay back the loan.

The practice of sharing a *goudguad*, which was discussed at the outset of this study (see Chapter 1) was not quite common in Ambassel in the 1984/85 crisis, but several informants reported that it was fairly widespread during the famine of the pre-revolution years. In fact, *goudguad* as a means of grain storage has become less important in this awraja in comparison with the surrounding awrajas of Yejju, Dessie Zuria, and Qallu, where more peasants use it to store food, and relatively more peasants participated in *goudguad* sharing in this last famine. We found only five peasants who said they had been involved in sharing a "prosperous" peasant's *goudguad*; two of them were in Tehulederae, and three in Worebabo woredas.

The literature on famine often argues that survival response frequently entails marked changes in the normal diet of peasant households. This may be true of some African and Asian experiences, but in the case of Wollo the changes that occurred were not so much in the diet itself as in the variety of the foodstuffs consumed, and the frequency of the meals served. As the food stock of the household is exhausted, the family is treated to a monotonous diet of one or two simple and poorly processed food items, often served only once a day. This regimen frequently indicates that a household has exhausted all its material resources, and all sources of support have also dried up. In Wollo, this simple diet consisted of *qollo* (roasted grain) or *niffro* (boiled grain). These food items are often made from any kind of cereals or pulses, the most common being barley, wheat, beans or chickpeas. Each family may be given only a handful (perhaps weighing not more than 50 grams) of one or the other of these foods at a specific hour in the day.

Both *qollo* and *niffro* are however popular light snacks in rural Ethiopia, and a regimen of one or the other does not involve a change of diet. In fact, even in stress-free times, *qollo* and *niffro* constitue the poor ploughman's lunch or supper. However, at this first stage of famine survival peasants attempt to supplement their diet with a variety of edible plants, roots, berries, wild vegetables and fruits—in short, famine foods—and this certainly involves a change in their diet and consumption habits. Famine foods, a subject to which we now turn, may on the other hand be seen as supplementary or complementary sources of food reserved, as it were, for extraordinary and critical conditions.

The traditional environmental science of the peasantry includes an extensive stock of botanical knowledge; the average experienced peasant knows a great many of the wild plants, trees, and natural forest resources in his/her local community. While both men and women are equally knowledgeable about the natural vegetation of their environment, the former specialize in plants which have medicinal, purgative, and constructional value, and the latter in those that may be employed as spices, herbs, food supplements, and decorative objects. It may be argued, in fact, that peasant women have more or better knowledge of edible wild plants than peasant men, though it is difficult to prove conclusively. The botanical knowledge of the peasantry includes identifying the properties, growing periods and habitats, uses, and similar other characteristics of each of the known plants all of which are distinguished by their individual local names[1].

The famine food plants that Ambassel peasants resorted to during the crisis may be grouped into three categories: (a) leafy plants and plants with seeds; (b) berries and "fruity" plants; and (c) roots. Let us begin with the last (which were the least common) and proceed to the first, which were relatively the most widely consumed famine foods in the awraja.

As was noted in our discussion earlier, Wollo peasants are not accustomed to cultivating drought tolerant root or tuber crops. One reason why crops like taro, yam or cassava are not popular may be because of their long maturation period which may last between eight and ten months. Some peasants in Worebabo were reported to have used taro as famine food; the plant is grown on a small-scale in this woreda. Peasant informants here also stated that the root of the banana tree was eaten by distressed peasants during the famine. It is also likely that some parts of the *koba*, a variety of the *ensete edule*, may have been eaten in the crisis; there is some indirect evidence to suggest that the pseudostem and the seeds of this plant may have served as food in both this and Tehulederae woredas. Both parts of the plant, and sometimes the shoots, have to be boiled to be eaten.

In the second category are berries and fruit-like plants. Generally speaking berries like *agam* (*clarissa edule*) and *quegga* (*rosa Abyssinica*) are consumed, particularly by youngsters, in normal times as a snack or a treat. In distress conditions these berries are systematically collected and consumed in the family. Another plant in this category which also serves as famine food is known in the awraja as *mekan's qoulqual*. It is a species of *qoulqual* (*euphorbia candelabrum*), but does not bear fruit (hence the term "mekan'e"), and is a close relative of the "prickly pear" (*opuntia vulgaris*). This plant is said to have dangerous side effects if consumed on a large-scale. Informants reported that a number of peasants had died of eating the plant. The dead were said to have been

[1] It is interesting that the Department of Biology's multimillion dollar project to identify and register all the useful plants in this country relies to a large extent on the expertise of peasants, and yet not enough credit has been given to the latter by local scientists and decision-makers for their knowledge. See, e.g., the reference under the Department.

severely dehydrated, and their bowel movements were blocked. A third plant in the same category is the *shola* tree, (*ficus sur*), whose fruit, and, occasionally, bark serve as distress food. The bark is stripped, and the inner, softer part is boiled and consumed.

In the first category are plants whose leaves and seeds are used as food. The most common of these are locally known as *antaria* (*portulaca oleracea L.*), and *aluma* (*amaranthus angustifolius*), both of which are found in many parts of Ambassel[1]. The plants *argagsa* (I was unable to find the botanical name of this plant), and *gourdo* (*Portulaca quadrifida L.*) were found mostly in Worebabo woreda, while *gougoublae* (*gugulbi* in both Mooney and Wolde Michael, *tragia brevipes*) was found only in Ambassel woreda.

In each instance, the leaves of the plant are boiled in water (sometimes, as in the case of *argagsa*, more than once to take out the bitter taste of the plant), and eaten either as vegetable, or prepared as sauce. The *aluma* plant has further uses: its seeds are ground into flour from which a thin, pancakelike bread is prepared. If the plants are collected when young both the leaves and the stem are cooked and consumed. Great care is taken in collecting these wild plants because there are other plants which closely resemble them but which are poisonous if taken as food. It is the responsibility of the women in the family to identify the edible plants and to select the useful parts; how and when each plant is prepared for consumption is also part of their stock of botanical knowledge.

Divestment and asset disposal

It was observed earlier that the average peasant routinely monitors the market in his locality to determine the movement of prices and of marketed products. In distress conditions, the peasant gives greater attention to the market as well as making greater use of it. This fact must be borne in mind in the discussion that follows. In fact, in these circumstances, the peasant's market horizon, as it were, expands considerably. He becomes concerned not just with the local market but with

[1] I have relied on Mooney (1963), and Wolde Michael Kelacha (1977) for the botanical names of most of the wild plants used as food. The plants antaria and gourdo were identified for us courtesy of the National Herbarium, Science Faculty, Addis Ababa University. Strelcyn (1973) was not useful for our purposes.

the markets in distant regions which he visits if he thinks there are advantages to be had.

The system of divestment adopted by peasants at the middle phase of famine varies from family to family, but there is one thing common in all cases, which is that the process is always selective and graduated. The standard argument on the subject holds that at a certain point in the distress cycle peasants sell their smaller stock first, which includes sheep, goats and often young calves. This is followed by young cattle, then cows, and finally work oxen (see, for e.g., Gryseels and Jutzi: 23; this is more or less the view of Holt and Cutler: 9–10). This system was certainly adopted by a number of peasants in Wollo and other areas in the northeast, but it would be too simplistic to consider this as a universal formula adopted by all or even a majority of peasants. This rather unpretentious model does not cover the wide variety of experiences of distressed peasants, although as a formal model it appears neat and logical.

What the formula forgets to take into account is that the value of an asset changes dramatically in the fluid conditions obtained in this phase of famine. It may be argued that as a general rule live assets considerably deteriorate in quality and hence in value, while "inanimate" assets (implements, household goods, fuelwood, building materials, etc.) lose less of their value in comparison as the crisis progresses. Indeed, some assets in this category, such as fuelwood, building material and the like, may remain stable for a considerable period of time.

Bearing in mind the discussion presented in Chapter 7, we suggest that the divestment plans adopted by a peasant household will be determined, or significantly influenced, by the following considerations:

(a) By the wealth of assets the household has at its disposal. Peasants with a greater variety of disposable assets will have a wider choice than poor peasants who are forced to sell their limited possessions at an earlier stage.

(b) The specific needs of the peasant family. For example, a family which has one or more members who depend on milk and milk products for their food may be reluctant to dispose of its milk-producing animal or animals, unless as a last resort. Take a close look at the livestock table, Annex 2, at the end of this study. We do not know accurately what the distribution of male and female cattle was among peasants in Ambassel before the famine, nevertheless, the fact that far more oxen and bulls were offered for sale than cows and heifer *all through the period* covered by the registry is not without meaning. If we assume an even distribution of male and female cattle among the population, the data suggests that far more people were willing to market male cattle, including their work oxen, than female cattle.

The third factor which has a bearing on divestment is (c) the special contractual or verbal arrangement a household makes with another household for the common use of assets, especially after the famine. We have already discussed this and we shall not add much here. The main point is that when two or more households agree to support each other, each household will adopt a different selective criteria for asset disposal than if it was acting alone.

(d) The connections—family, kinship or religious—a household has with one or more households in stress-free areas, and how much assistance it receives or expects to receive. A family with a strong connection in famine-free areas may acquire donations (of food or cash), loans, and a "safe haven" as it were for its live assets. We shall return to this last point further down.

Finally, (e) the demands of the market, and the behaviour of prices. While it is true in general that the price of live assets falls progressively and that of food rises in the same manner, there may be short-term changes or reverses in both instances from time to time caused by extraneous factors. Some peasants may thus try to take advantage of such reverses, which may involve rearranging their original divestment plans. Look again at Annex 2. In May 1984 there was heavy rainfall in many parts of Wollo (see Annex 1), and the number of cattle offered for sale was the lowest for the period covered in the table; prices also remained stable, except for calves in this month.

In this section we shall be concerned about livestock disposal and livestock flows during the famine, but before we get to that subject a few words about asset disposal in general.

An asset is any item of possession, or any resource that may be sold in the market or exchanged for food through barter. Thus everything a peasant family owns, including the members' personal effects, are marketable and marketed. There are two exceptions to this, the one prohibited by law and the other by moral convention, and these are land and people. The land reform of 1975 does not allow peasants to sell land or to exchange it for other property or assets. Contrary to some allegations (see Cutler and Stephenson: 6) people were not sold into slavery during the recent famine, nor the one before that. We shall return to this point later.

As was pointed out earlier, a good number of peasants sold their dwellings at the high point of the crisis. This was a last resort measure, but houses (or rather the materials they were built of) fetched relatively better prices than livestock whose prices were severely depressed at this point. Houses were sold partly as fuelwood, and partly as building material, and the prices of both items were fairly stable in Wollo in the period under discussion.

The sale of personal effects, including jewelry, by peasant women has been seen by some (e.g. Jiggins 1986) as evidence of the heightened exploitation of women during a food crisis. A good number of the women we interviewed in Wollo said they had sold all or most of their personal belongings: jewelry, clothing, etc. One woman told my field assistant that she sold everything she personally owned, including all her clothes, except the ones she was wearing, before she found a relief camp and was admitted to it.

As a general rule, women in nomadic cultures invest more on jewelry and adornments than women in sedentary farming systems. This is a form of saving, and it is recognized as such by pastoral societies, which are subjected to more frequent crises than settled agricultural societies. At the time of the famine, many markets in or close to pastoral settlements in Wollo were flooded with jewelry and women's personal effects which were often sold at very low prices. But these same markets also offered a glut of men's personal effects, including a large supply of hand weapons of all kinds. To the pastoralist man, his hand weapon—a knife, dagger or short sword—is one of his most prized possessions, and the chief symbol of his manhood. In brief, what we are proposing is this: the sale of women's personal belongings in times of crisis does not necessarily indicate increased exploitation, but may be, at least in a good number of cases, part of the divestment plans of the family.

There is one other aspect of "asset disposal" which we should mention before moving on to discuss livestock marketing. Peasants with relations or friends in distress-free rural communities often attempt to take advantage of the situation through a variety of arrangements. One of the most common was an arrangement in which a distressed peasant would transfer some of his live assets for safekeeping with his relatives or friends there. Several peasant informants stated that they had moved their oxen, a few their cows, to other areas during the worst months of 1984. One young PA leader, the secretary of PA No. 05 in Ambassel woreda informed us that he took his young bride, a cow and a few goats—his main assets—to his parents, who live in the higher *degga* area of the same woreda, and who were not affected by drought; he then left for Addis Ababa to work as a daily labourer. He returned home in the latter part of 1985 and collected his wife and his livestock.

These arrangements may sometimes involve remuneration, particularly if the two parties are not close relations, and if the arrangement is expected to last for an extended period of time. A peasant in distress may place his livestock with someone for safekeeping with the understanding that the latter will keep for himself one offspring of one of the animals—i.e. a kid, ewe or calf, as the case may be. This is in recompense for feeding and caring for the animals.

Here are five cases, taken from all three woredas of Ambassel, showing how complicated the divestment plans adopted by peasants were during the famine. Since the drought had been going on for at least more than a year, we don't know at what point each peasant decided to dispose of his assets in earnest (compare the cases with the data in Annex 2).

Case 1: This peasant had several sheep and goats (including two *moukit*, i.e. fattened sheep or goat), two donkeys, one cow and one ox. His case fits in very well with the standard model of asset disposal noted above. He said first he sold his *moukit* sheep for 50 Birr[1], and a short while later he sold a *moukit* goat for 35 Birr. This was followed by one donkey which he sold for 23 Birr (this must have been in the summer of 1984). Next he sold his other donkey for 27 Birr. His final sale was his cow which he got rid of for 35 Birr, most probably during August or September (see Annex 2). He was able to survive the crisis with his one ox.

Case 2: This peasant is from Ambassel woreda (PA No. 07). It was not unusual for peasants to indulge in a little bit of speculative livestock buying and selling during the famine. The peasant said he first sold his cow and bought a heifer (no prices were given). He then sold his donkey to buy food. Next he sold one of his two oxen and bought a young bull, and finally he sold his last ox. At the end of the crisis he was left with one heifer and one bull.

[1] He must have started selling live assets in the spring of 1984 as this kind of price for sheep was not uncommon at this point. This peasant was from Ambassel woreda, PA No. 05. But he was still in distress, so he sold his house, which was not too far from Bistima, the awraja capital, and moved into a small makeshift hut on his farm plot.

Case 3: This peasant is in Worebabo woreda, and from what he told us he was not a poor peasant before the famine. He then had a heifer, a cow, two oxen, two donkeys and a tin-roofed house. He disposed of his assets as follows: first he sold his two donkeys in turns, then his heifer, then the cow, and finally his two oxen.

Case 4: This peasant is also from Worebabo. He begins by selling his calves (we forgot to record how many calves he had), then goes on to sell his donkey, then his ox, and finally his cow (he owned one each of these animals). This was not sufficient, he said, to see him through the crisis, so he sold his farm implements (the plough and all the accessories). He still was in serious trouble, so he sold his dwelling as a last measure.

Case 5: This peasant is from Tehulederae; he had two calves, one cow and one ox. His divestment measures took the following form: first the calves were sold one by one, then the ox and finally the cow. At the end of the famine he had no animals in his possession. At the time of the interview he had bought one female sheep.

Many peasants in Wollo do not consider disposing of small stock as distress selling. Small stock, and particularly male goats, are often marketed in normal, stress-free times to raise cash. If small stock are marketed for purposes of coping with crisis, young, male sheep are sold off first. Female small stock are sold after males. Some peasants prefer to keep goats well into the high phase of the famine because these animals withstand stress better than other animals, small or large.

Here is a rather unusual example of live asset disposal reported by the provincial office of the Ministry of Agriculture in Wollo (MOA, Wollo 1985: 68–71). According to a document of this office, some peasants disposed of their cattle at an early stage of their divestment activities, and tried to retain their small stock to the end. The rationale for this is that small stock require less food than cattle, and in case of death less loss is incurred. The loss of an ox or a cow means the loss of a large investment. If the small stock survive through the crisis, they are sold to buy young cattle.

Livestock flows during the famine

Map 2 (p. 67) attempts to show the movement of livestock out of the famine zone in 1984/85, with Ambassel awraja taken as the centre of the flow.

The loss of livestock suffered by the peasantry in Wollo in the period

1983 to 1985 is quite considerable, although the exact number will never be known. A great number of our informants in the province stated that they had sold some or all of their livestock during the drought. An equally large number said they had lost one or more livestock through death. Almost all extension agents, however, stressed that far more farm animals were sold than were lost through death by starvation. In the fall of 1984, the prices of cattle reached rock bottom, but peasants were still eager to sell, and large volumes of animals were brought to market. Interestingly enough the price of equines was relatively more stable than that of cattle or small stock, and fewer of these animals died of starvation. One reason why death toll of livestock was less than expected was because of the sporadic showers that fell periodically during the drought.

In this discussion we shall speak of livestock "losses" in the absolute and relative sense. Absolute loss involves death of livestock while relative loss means loss by sale. More on this further down.

The Ministry of Agriculture office in Ambassel tried to assess the number of cattle lost in the awraja through death and sale in the Ethiopian year 1977 (1984/85 G.C.). The inventory was compiled through the PAs of each woreda; each PA was asked to record the number of animals a peasant household had lost in this manner. The result was a collection of figures that most extension agents in the area now believe is highly inflated. Peasants must have thought the government was about to distribute farm animals when the inventory was taken. For whatever it is worth, the data is shown in Table 24. Some seven-tenths of the equines sold, and three-fifths of those which died were donkeys. Surprisingly enough, the inventory shows that Ambassel woreda, which was the least affected by the drought in the awraja, lost more livestock through sale *and* deaths than the other two woredas.

The evidence available in the provincial office of the Ministry of Agriculture in Dessie, on the other hand, suggests a somewhat different picture. According to this office, about 35,000 heads of cattle were "exported" from Wollo to other areas in the country in the first six months of 1977 E.C. (1984/85 G.C.). Some 25,000 of these cattle were driven to Gojjam and Gonder (through Lasta awraja in Wollo) by merchants from these two provinces, and another 9,100 were driven south to Addis Ababa, mostly from central Wollo (MOA, Wollo: 68–71).

A few words are in order about livestock and livestock marketing in

Table 24. *Livestock sales and deaths in Ambassel awraja, 1984/85*

Livestock	Total number	
	Sold	Died
Oxen	20 231	13 478
Cows	16 376	11 479
Heifers	13 831	11 279
Bulls	8 826	5 038
Calves	7 840	5 560
Sheep	9 931	10 004
Goats	13 919	3 473
Equines (ex-camels)	10 457	5 105

Source: From the records of the Livestock and Livestock Products Marketing Unit, Ministry of Agriculture, Ambassel awraja, Haiq, October 1986.

general in Wollo. This province has a rich resource in its farm animals, even though large numbers of them are lost in the periodic famines that afflict the area (see Table 8). According to a study conducted by the Australian Agricultural Consulting and Management agency (AACM), Wollo has the third largest number of cattle in the country, behind Shoa and Sidamo; in terms of livestock markets, it is also the third largest, bettered only by Harar and Shoa (AACM, Annex 10: 8, 15; this is a valuable study for those interested in livestock marketing in Ethiopia).

We may distinguish three levels of livestock markets in Wollo: the feeder market, the area market, and the terminal market. The feeder markets are those located near the main trunk road, or along the main livestock routes. The main feeder markets in Ambassel are Golbo, north of the capital Haiq, Adami to the east, and Gerba in the southeast. In south central Wollo, Albuko in Qallu awraja, Jama in Wore-Illu awraja and Kelala in Borena are important feeder markets and also key transit points in the southerly flow of livestock. North of Ambassel, we may cite Girana as an important feeder market; in western Lasta, Filaqit serves as a transit point for the westerly flow of cattle (see Map 2).

Haiq, the capital of our awraja, is an important area market, so too Woldiya in Yejju. There are two kinds of terminal markets, local and national. The local terminal markets in Wollo are Dessie and Kombolcha. Here, a large number of the cattle marketed are destined for the slaughter-houses, and the meat packing factory in the case of Kombolcha. The rest are driven to the national terminal markets of Addis

Ababa, Asmara and Assab. Those cattle driven to Assab are exported overseas.

In the feeder markets, the peasant sells his livestock mostly to another peasant; however, perhaps a good third of the livestock marketed here are bought by what we call rural livestock traders, the overwhelming majority of whom are peasants. In the area markets peasants sell livestock to rural traders as well as to large-scale urban-based livestock merchants who also buy from rural traders. In all the terminal markets, merchants are the most dominant force.

In the famine years of 1984/85 the number of livestock marketed was exceptionally high. Who were the buyers and where were the livestock taken to? To find out the identity of the buyers we asked our respondents who they thought were the most active livestock purchasers in the markets that they were familiar with. Their answers are given in Table 25.

If we leave out the "unknown" column we find that 55 per cent and 37 per cent of livestock buyers in Ambassel during the famine were identified by peasants as merchants and peasant traders respectively. Actually, if we add to this information supplied by extension agents and PA leaders the picture becomes a bit more complicated. The main livestock purchasers in the period under discussion were the following:

1. Licensed livestock merchants. These bought livestock mostly for terminal markets. Some of the "unknown" in Table 25 may have been commissioned livestock agents working for the large merchants. The most active merchants during the famine (apart from the local ones) were those from the provinces of Gondar, Gojjam and Shoa. The livestock flow north involved merchants from Raya and Qobbo.

Table 25. *Identity of livestock buyers according to respondents, Ambassel, 1984/85*

Buyers	Number of respondents who identified			
	Ambassel	Worebabo	Tehule.	Total
Cattle merchants	23	32	28	83
Peasant traders	25	21	10	56
Urban dwellers	5	1	1	7
Unknown[a]	11	45	49	105
Other	1	1	2	4

Note: [a] Peasants did not know the identity of buyers.

2a. Peasants and peasant traders. The most active peasant livestock traders in Ambassel came from Borena, and Wore-Illu awrajas, with a small minority from Wore-Himenu awraja. These traders preferred young cattle—bulls, heifers, and also calves—instead of sheep and goats. They sold their flocks in area or local terminal markets to the large merchants or took them to drought-free areas in their awrajas.

2b. On a smaller scale, ordinary peasants who could afford to make limited investments, and who were not themselves in distress, bought a small number of animals, particularly when prices were low.

3. Urban businessmen and speculators: These were mainly interested in acquiring cattle and shoats for local consumption, and only a small amount of the animals marketed were bought by these individuals.

No agency of *government was involved in purchasing livestock from the peasantry*.

We noted earlier that sale of livestock by the peasantry involved a relative loss; we will now discuss what we mean by this. Most of the cattle, and small stock marketed in Wollo left the province for the terminal markets, but a good number remained inside. Of those which were driven out of Wollo, some went to the slaughter-houses of the major cities, especially Addis Ababa and Asmara. Since the demand for meat among the population in these cities is not very elastic, very little excess livestock was absorbed in this way. In the second place, a good proportion of Wollo cattle were sent for export through Assab, and also Massawa. Some of the excess animals marketed may have found their way into the two ports, but since these can only handle a limited volume of livestock, the excess must have been modest. (See AACM, ibid, Ch. 2.4 for the limitation on exports.)

It is our opinion that a large percentage of the excess livestock marketed in Wollo and the northeast during 1984 and 1985 was illegally exported to the neighbouring countries of Sudan and Djibouti. The AACM study noted above (ibid: 25–26) reports that the illegal export of animals to these countries, and also to Somalia and Kenya has been going on for some time, and was estimated for one year in the early 1980s to involve a quarter of a million heads of cattle. The "underground" exporters of cattle to the Sudan in our case were mostly merchants from Gondar, who bought their animals in Woldiya and the western awrajas of Wollo; some Gojjam merchants may also have been involved in this work on a smaller scale. Illegal exports to Djibouti took place along the Bati–Assab road, and the exporters here may have been Wollo and Shoa merchants. There is some indirect evidence that some Wollo cattle found their way into Harar province.

Livestock sold by distressed peasants flowed to the various destinations through what may be called a system of relay trekking. At every major transit point in each direction, the ownership of a herd changes hands. Let us look at the southwest route in Map 2. Wore-Illu and Borena rural traders drive their flocks from eastern and central Wollo to the transit points of Kelala and Jama on the Shoa border. They sell their animals to Shoa merchants and rural traders, and these then drive them through Shoa, and sell them to merchants in the capital. In the same way Raya Qobbo merchants purchased livestock from traders and peasants in Girana and Woldiya for the trek north.[1] Gondar and Gojjam merchants may pick up animals from smaller traders in Woldiya, or from the important transit point of Filaqit in Lasta for the trek west.

The foregoing discussion concerns livestock taken out of the province for good. What happened to those which were driven out of the famine zone only, and which ended up in the drought-free areas of Wollo and adjoining provinces? There is sufficient reason to believe that a good percentage of the livestock in this category returned to Wollo at the end of the famine. The same cattle routes were used to bring back the animals to the various feeder and area markets for sale to the peasantry, mostly by peasant livestock traders. In this period of recovery, small peasants in drought-free pockets in the famine awrajas, who had bought livestock during the drought, brought them to the market to sell. It is very difficult to estimate the number of animals that came back to the post-famine zones, but our guess-estimate for Ambassel is that something in the neighbourhood of 20 to 25 per cent of the animals lost to the peasantry of the awraja through distress sales returned in the early phase of recovery and were marketed there. The process of reverse livestock flow was still going on at the time of our field-work.

Several peasants in several parts of Ambassel, who had lost oxen in the drought, said they were able to buy oxen at the end of the famine. A peasant in Worebabo woreda told my field assistant that he bought back the same ox he had sold a year earlier from a peasant trader. This is of

[1] There was a brisk trade in equines, especially donkeys in 1984/85. Most of these animals were driven to the salt mines in northeast Tigrai.

course a rare coincidence, but the point is that peasants were not deprived of access to livestock at the end of the crisis thanks to the rural market system, and the rural peasant trader. This individual provided an important service to the peasantry, first by buying farm animals and taking them out of the famine zone (i.e. saving them from death), and second by bringing them back to sell to the people who needed them in the period of post-famine recovery.

It is interesting that this simple but effective technique of buying up livestock from the peasantry in times of food crisis, taking them out of the famine zone, and bringing them back to sell to the peasantry at the end of the crisis—a task the average peasant has been doing for decades—has never been considered as a famine response measure by government at the national or local level.

Normal and abnormal behaviour

This may be the right place to discuss this subject, although it deserves far more extensive treatment than we can provide here. The subject has been treated lightly or in a cavalier fashion in some of the literature, and this has given rise to a large variety of unfounded beliefs and absurd and fanciful opinions. Some writers, who one would think should have known better, have contributed to these fantastic views by failing to treat oral tradition having to do with the subject carefully and critically, and by accepting what obviously are apocryphal stories as scientifically valid.

The first thing that reasoned analysis should attempt to clarify are the terms "normal" and "abnormal". Because of space limitations, we shall not enter into a sociological and psychological discussion of these two concepts, nevertheless, the question, what constitutes normal or abnormal behaviour in a social situation which is itself highly abnormal, should always be soberly considered if this subject is to be treated seriously and objectively. Moreover, whatever evidence is available in individual cases must at all times be carefully weighed and examined.

Among the large number of acts of abnormal behaviour said to have been carried out by peasants in famine conditions, the following are the ones which have been treated in the literature with far less objectivity, and far more sensationalism than they deserve: child abuse, abandonment of children, and inhuman treatment of the weaker members of the family by the stronger; entering or selling oneself or others into slavery;

murder by negligence, or through selfish acts; high incidence of criminal behaviour, mainly robbery and murder; and cannibalism, particularly child canibalism by mothers (Pankhurst 1985, Mengistu Lemma 1959 E.C., Cutler and Stephenson). In the pages that follow we shall briefly comment on each of these alleged acts of abnormal behaviour in connection with the recent Wollo famine.

The basic pattern of normal behaviour, says a report (Cutler and Stephenson: 6):

> ...is supplemented by the abnormal social behaviour typical of famines. As a last resort people will desert their families and sell their bodies for prostitution or slave labour. Children will be driven away by their own parents and old people will be abandoned to their fate. This kind of behaviour is currently occuring in parts of Ethiopia, and has happened despite the existence of both an early warning system and a relief structure unique in Africa.

Now, this is nonsense, and we shall attempt to show why. Let us begin with prostitution and slave labour. There has been (and still is) a steady and continuous flow of rural women to the cities ever since modern urbanization started in this country more than a century ago. A good number of these women end up as bar girls and "women of ill-repute" because this is the only form of employment available for most immigrant women; others find work as domestic servants, and provisioners of traditional food and drinks. A famished woman will have no suitors, and will not be accepted as a prostitute by her employers, and rural women (especially the young and healthy looking ones) migrate to the urban areas for employment before famine has actually struck, particularly at a time of prolonged pre-disaster difficulties. In many of the bigger towns, like Dessie or Kombolcha in Wollo, entry into prostitution is not as easy as it may appear; rural women planning to enter the profession have to compete with likeminded urban women, who are often more aggressive and more "equipped" for this kind of work. An urban girl enters the profession not in the town where she lives or has many relations, but in a distant place, often in a different province. There is thus a steady inter-urban flow of town girls seeking, like the rural women, easy and non-skilled employment.

We found no evidence not even one case where men, women, or children had been sold into slavery, or made to do "slave-labour" (whatever that means) in Wollo during the famine. It is also highly unlikely that anything of this sort took place in the region as a whole during the previous famine. The rural areas have undergone a serious economic impoverishment, and very few peasants can afford "slave-owning", which is in reality a burden and an additional mouth to feed. The economic decline of the urban areas is equally serious, and very few opportunities exist for hiring peasant "slave-labour". Besides, urban labour is dirt-cheap, and a prospective employer is better off with this than with peasant labour.

There is one practice in Wollo, and in several areas in the northeast, which the authors under discussion were obviously not familiar with, but which may be taken as an indicator of the deep poverty that exists in the rural areas. This is known as *g'it'ir,* which may be roughly translated as *child contracting*. This is a system in which a well-to-do peasant contracts a youngster from a poor family to live in his house and work for him. The job is usually livestock herding, but it could also involve light farm work. Contracting peasants usually prefer boys, aged ten and above and the parents of such children receive an agreed upon sum of money periodically, in many cases about 30 Birr per year. On occasions, children may be contracted by townspeople to do odd jobs for men.

Now, the interesting thing is that peasant "contractors" affected by famine return the children to their parents to reduce the number of mouths they have to feed, and may accept them back in the post-famine period if they (the "contractors") have not suffered seriously in the disaster. Poor peasants involved in child contracting thus find temselves more burdened during famine than before it, and "abnormal" hiring practices cease to function during times of crisis.

A word about the abandonment of "old people to their fate" which is more difficult to prove or disprove. If this was a widespread practice—like "child abandonment", discussed below—we would have known more about it. The composition of inmates in the relief camps, particularly, the age structure of such populations would have been useful as evidence of what happened to the elderly in the famine, but this kind of information is hard to come by. One of the first shelter camps to be built in Wollo was the one put up by the Save the Children Fund (SCF) in Korem in early 1983. An eyewitness account reports that most of the peasants who flocked to the town to seek shelter and relief were women and older people, the younger men who were conspicuous by their absence, having either stayed in the rural areas, been commandeered for service by anti-government groups operating in the area, or gone elsewhere to find work or assistance (see "Anonymous" 1983). A report submitted to the Ministry of Health in Wollo by SCF shows that of the total deaths that occurred in the Korem shelter between September 1984 and August 1985, 11 per cent were men and women aged 65 and over. By this time shelter camps in this area were showing a more even distribution of the population, but the elderly were "more represented" in the death figures. In short, there is no conclusive evidence that people who were advanced in years were abandoned to their fate during the famine.

The subject of child abandonment is a complicated and contentious one. Let us see if we can untangle the web of prejudice and unfounded allegations surrounding this subject. The word "abandon" is a misleading one, and needs to be defined carefully if the subject is to be discussed seriously. There are two forms of "abandonment" and each is accomplished by a motive entirely contrary to the other. In the one case the motive is to let the child die of hunger or of the elements, in the other, to try to save it.

A mother (it could also be a father) dumps her child somewhere where there is no help to be had and walks away knowing that the child will die, or some terrible harm will come to it. She may leave it in the house and go away with similar intentions. This is the real meaning of the term child abandonment. On the other hand, a mother, who senses that both she and her child will soon die of hunger, will try to force the baby on someone (a passerby, a stranger, etc.), or leave it at somebody's door and go away hoping that this someone, or the owners of the house will take pity on the child and save it from death. This is not abandonment, but rather the desperate attempt of a mother to give her child the chance of life. We cannot speak of these two acts as being one and the same: the first is child abandonment with ill intentions, the second abandonment as a survival measure of last resort.

In reality, what happened in Wollo was far more complicated than that, nevertheless, we have no evidence that child abandonment in the first meaning of the word, did or did not take place in 1984 or 1985. The information available with NGOs responsible for abandoned children in these years refer to abandonment mostly in the second sense of the term, although the other meaning could also be imputed to some of the parental behaviour. It is our opinion that if child abandonment with ill intentions did take place in Wollo, it must have been on a fairly modest scale.

This writer was told by eyewitnesses of several incidents in Wollo where mothers tried to shove their children into the arms of travellers and strangers at the height of the famine. A research team from this University, doing survey work in Yejju awraja just north of Ambassel in 1984, decided to adopt a young peasant girl because her mother, thinking both of them had very little chance of survival, took a gamble and forcibly directed the girl into the inn where the team was staying, and almost into the arms of several of the team members.[1]

According to a number of local extension agents working for NGOs responsible for abandoned children, or involved in emergency children's centres in Wollo and north Shoa there were several reasons why and how children were separated from their parents, or found themselves "abandoned":

1. Parents and children travel to the relief centres for assistance, and there the parents die of famine-related causes.

2. In severe cases, children are separated from their parents and are cared for in children's wards in the shelters. The parents may decide to leave their children in these centres and return home in the belief that they are better off there (which is not untrue). The children are reunited with their parents later, and this is still going on today.

3. There is the reverse case of children abandoning their parents. According to one relief agent, quite a few of the cases he has handled in Wollo were of this type. This happens particularly if during a high point in the crisis, children believe they are being a burden on their parents, or if there has been ongoing friction between a child and a parent. What the children do in this event is walk to the children's centres or relief camps and ask for assistance. In a number of instances, children may just run away from home, and may be found wandering about by someone. As was observed earlier, famine does exacerbate conflicts within the family and between parents and children.

4. Children were separated from their parents during resettlement activities in the northeast. This may be because the parents had died while awaiting relocation, or children were separated in the confusion and chaos that accompanied the work of transporting peasants to the resettlement sites.

According to one source there were about 2,600 unaccompanied children in Wollo in 1986, of which maybe about one-third were orphans. Earlier, however, the number was quite high, as shown in Table 26.

[1] Yeraswork Admassie, Sociology Department, oral communication.

Table 26. *Unaccompanied children in Wollo (by awraja) (in shelters in the period Nov. 1984–Sept. 1985)*

Awraja	No. in shelters	No. of child deaths
Qallu	1120	11 627
Aussa	130	141
Ambassel	732	341
Wag	447	204
Raya Qobbo	454	499
Yejju	127	160
Lasta	107	NA
Dessie Zuria	1364	43
Wore-Himenu	225	211
Wadla Delanta	170	NA
Borena	None	None
Wore-Illu	None	None
Total	4876	13 226

Note: NA = not available.
Source: NEERNDRC 1985 A:19.

The most important local NGO which was the largest recipient of unaccompanied or abandoned children was the Ethiopian Orthodox Church (EOC). The number of children from the famine areas of the northeast and adjoining localities received by the Church's Children's Council in 1985 is given in Table 27.

Table 27. *Children received by EOC from famine areas, Northeast, 1985*

Province	No. of Children
Wollo	1010
Tigrai	472
Gondar	89
Gojjam	14
Total	1585

Source: Ethiopian Orthodox Church 1986:18.

Let us now move on to the question of cannibalism, and child cannibalism, a practice alleged to have been indulged in not just by famine victims in Ethiopia, but also by others elsewhere. For example, a work by two western Marxists states that cases of cannibalism were reported by the Soviet Press in the 1921 Russian famine (Hussain and Tribe

1983: 267). In our own case, the work which has sensationalized the subject is the one by Pankhurst noted earlier (1985: 82ff). Pankhurst has done his readers a disservice for having failed to sift through his evidence carefully, and for taking all allegations of his sources at face value.

His report concerns the great famine of 1889–92, in which, we are told, "unnatural" practices such as sale of children by their parents, self-enslavement, murder and cannibalism were widespread. He speaks of many cases of mothers eating their own children. He reports of a case where a woman from Ensaro in today's North Shoa is brought before Emperor Menelik and when asked what she had done, confesses to having eaten seven children. The story comes from the Emperor's Chronicle. The imperial chronicler, Gebre Sellassie, speaks of people eating each other, and of a woman in Wollo who ate her own child during the high point of the famine. However, he does not provide credible evidence for these stories which obviously are based on hearsay.

Another historian of Menelik's period also refers to the same story about the woman cannibal (Afeworq 1901: 63). Afeworq was not present when the woman told her story, and no one actually saw her killing and consuming human flesh. It is not uncommon in this country for deviants, mystics, or people who dabble in the occult—and the woman brought before Menelik must have been one of these—to claim to have carried out extreme or strange acts, or to know of fantastic things; this is to draw the attention of men of power or wealth, and to win favours from them. This particular woman's prize was a stock of food which she was granted by the Emperor.

Aleqa Lemma, a traditional scholar, also speaks of widespread cases of child cannibalism during this same famine, which he lived through. "There were many women", he says, "who killed and ate their small children." Furthermore, "men ate leaves; there were no leaves that were not eaten. I do not know of any leaves that were left untouched" (Mengistu Lemma: 143). He further alleges that some refugees from famine who were camped in the newly established capital, Addis Ababa, kidnapped several children and ate them. Aleqa Lemma may have embellished his stories because at the time and in the years since, cannibalism was featured in all oral accounts having to do with this or subsequent food crisis.

No one has yet alleged that men ate human flesh in Wollo or the northeast during the last two major famines, but such is the climate of opinion in this country on this issue that many would readily believe it if serious allegations were actually made.

We believe all of the stories about cannibalism, especially about child cannibalism reported by the works noted above, or by others, are either spurious or based on unconvincing evidence. Pankhurst, for instance, does not ask how the woman managed to eat seven children over a period of time without arousing the objections of the community, or of the parents of the unfortunate children. If this woman had actually done what she claimed to have done even to one child, she would have been lynched or stoned to death by irate parents. Similarly, Aleqa Lemma does not say anything about the reaction of parents whose children are kidnapped and consumed. He himself did not see anyone consuming human flesh, nor interview anyone claiming to have done so. It is also curious that when everybody else is suffering hunger, Aleqa Lemma's child victims are plumb and healthy.

The most unsavoury and repellent act an individual can do is consume the flesh of another individual. This is the more so if the same individual has been brought up in a moral culture strongly influenced by the great religions, such as Christianity or Islam, which was the case of famine victims in northeast Ethiopia. It is a well-known fact that children are the most vulnerable and easily victimized group in society when famine strikes. The first thing that happens during hunger is that children become highly emaciated. By the time men and women reach the point of desperation and begin to entertain the idea of consuming their own children—this might be six or eight weeks after serious hunger has set in—large numbers of the child population will have died, and those remaining will have been reduced to ghostly skeletons. There is no flesh on a child which has starved for many weeks.

Those who are quick to raise the specter of "abnormal behaviour" conveniently forget that food taboos exercise a strong influence on peasant consumption and dietary habits. Among the rural population in northeast Ethiopia only a limited variety of wild game is eaten as food. So strong is the taboo against proscribed animals—including domestic animals such as donkeys, mules and horses, or bush animals such as monkeys, foxes, rats, and lizards—that people would rather starve than consume prohibited flesh. Needless to say, human flesh carries with it the strongest taboo. Peasants in the northeast are in fact less liberal in their dietary habits than peasants in the lowland areas of the country who have a relatively wider selection of forest resources to use as food in times of hunger. Monkeys, rats, hippopotami and crocodiles are readily consumed by hunters, shifting cultivators and semi-nomadic peoples in the marginal areas of the country; none of these animals are however considered fit for human consumption among peasants in Wollo or the northeast.

We may have belaboured this issue, but the point we wish to make needs to be emphasized. Any report about abnormal behaviour among famine victims must be based on solid investigation and analytically valid evidence, or else should not be made at all. The least an honourable writer can do is not burden the destitute, the starving and the dead with spurious allegations or disreputable characterizations.

We found no evidence that Wollo peasants had indulged in "unnatural" acts in the recent famine. Men and women had sold their possessions, eaten wild plants, gone about begging with ghostly looking children on their backs, and stayed hungry for days, but there was no incidence of child abuse, slavery, murder or cannibalism. A few instances of "mercy killings" are known to have occurred in some shelter camps; children who had gone beyond the critical threshold but whose death was somehow drawn out were "helped along" by agonizing parents to bring to an end their suffering. However, such acts were rare and it would be absurd to include them as part of the abnormal behaviour of famine victims.

It is likely that before communities were hit by the drought, some parents may have attempted to have their children placed in *g'it'ir* with prosperous peasants or townspeople. It is also possible that some peasants may have contracted themselves to others in return for food or cash loans; the practice here involves promising a certain number of days of labour for a certain amount of food or cash, and the difference between this and a standard credit arrangement is that payment involves labour. There is nothing abnormal about this or child contracting as it occurs quite often under normal circumstances.

There is some evidence that "suicide" may have occurred in certain parts of Wollo. The team of University researchers mentioned earlier report of a term they came across on several occasions in Yejju awraja. Peasants, they state, used the term "beroun zegtual", i.e. "he has closed his doors" to refer to someone who has committed "suicide". A peasant who believes that he has no alternative but death goes home, locks his house, and lies down to die; he may occasionally do it with the whole family (Yeraswork and Solomon 1985: 14–15; also, oral communication, Yeraswork). This may not be considered "suicide" in the standard meaning of the term. To the peasant who carries it out, it is a dignified way of dying.

We came across more or less the same term in central Ambassel, but here it referred to something different. A peasant here who "closes his doors" is one who is suffering from *social shame*. Such a peasant has taken food when everybody else was going hungry, and he had done that in hiding, behind closed doors. As Mesfin has pointed out in a paper (September 1985) the ethic of cooperation, popularized in such traditional sayings as "he who eats alone, dies alone" continues to have a strong hold on the peasantry even in times of crisis.

9. Post-Famine Recovery

When famine finally returns, as it were, to its hide-out beyond the rugged mountains, it leaves behind desolation and despair. People as well as nature are thoroughly exhausted, and the signs of death and devastation are there for everyone to see. Human survivers are drained of energy and the will to think, and while some may take between three and six months to recover, others may be impaired for the rest of their lives. Many households will have gone through drastic changes, and many will have lost—some partially, others completely—their key economic asset, namely family labour power. The earth itself is hard and resistant, making farm work for men and their equally exhausted farm animals doubly difficult.

The immediate post-famine period is a period of household adjustments. Families which have lost one or more members, particularly those which are deprived of one of the spouses and/or able-bodied persons, are at a distinct disadvantage, and will have to make up for the loss in one way or another. Unattached women, and women whose husbands have died or are missing are greatly sought after by men who have lost their wives during the crisis. The period is thus a period of marriages, or rather remarriages, and of family restructuring in general. Widowed men are more eager to remarry quickly than women, not only because they need more hands in the field but also because they cannot cope without women's household food management skills. For widowed women, on the other hand, remarriages often means more labour and ever more drudgery, but it may bring with it greater security too, at least for those in their twenties and thirties whose children are not old enough to assume family responsibilities. However, a good number of women may decide to stay independent and manage their own homes and farms by themselves with the help of their families.

According to the registers of the WPAs of the three woredas of Ambassel, a considerable percentage of rural households were headed by women in the early stages of post-famine recovery. In Worebabo woreda, the worst hit of the three woredas, 22 per cent of the household heads listed in PA rosters were women, and in the two other woredas the figure was 17 per cent. Not all these women were old or of unmarriageable age; some, who could have remarried if they wished to, chose to stay independent for a variety of psychological and economic reasons. Women who can rely on their own family labour, or who can easily acquire outside support, or arrange mutually beneficial farming set-ups are less likely to remarry if widowed than women in other circumstances. Widowed men, however, have no choice but to get new wives for the reasons given above, unless the burden of domestic labour is shouldered for them by female relatives or kinfolk.[1] While in the broader social and economic context women play a subordinate role in rural Ethiopia, men are dependent on them for their domestic labour and skills, and this fact has not been given sufficient weight in the debate on the subject.

The obstacles to full recovery are many and varied, but the most important constraints in Wollo in the autumn of 1985 were the following: lack of draught animal power, lack of seeds for planting (many had eaten their seeds during the famine), poor personal health of family members due to prolonged malnutrition, lack of sufficient family labour, poor condition of livestock, and poor condition of farm plots. Surviving livestock need sufficient feed and water to regain weight, but how quickly this can take place depends on the behaviour of the rains. Some have argued that cattle adapt to drought conditions fairly easily, and will quickly regain weight and health when the rains come (Topps), but this depends on how quickly feed is available. It has often been the case that severe drought has been followed by too much rain and flooding —indeed, this was what happened in Wollo in the autumn of 1985—which may reduce pasturage and the amount of food available to animals. For small stock, the problem in Wollo was predators, particularly hyenas and foxes, which reappeared after the famine and were very aggressive.

Those without or with reduced draught power will have to live with that disability for anywhere between two and three years, unless they have access to outside support. This disability, however, is not fatal, as

[1] For some idea of the burden of domestic labour having to do with food preparation in Ethiopia, see Selinus 1971.

we shall see further down, but it will continue to have a debilitating effect on a household's production and income (see Toulmin 1985: Ch. 3, Gryseels 1983). On the other hand, peasants will take a variety of measures to overcome or partially reduce the adverse effects of the lack of draught animals.

In the first years of recovery in Wollo—recovery for some began in the autumn of 1985, for others in the spring or autumn of 1986—single ox owning peasants teamed their animals with donkeys for farm work. In Wollo, and many parts of the northeast, equines are not normally used as draught animals. Many peasants in this same condition had also tried teaming a cow with an ox for ploughing purposes, another innovation or hardship measure brought on by post-crisis necessity. "*Lamituan gerahuat*"—I trained the cow to farm—was an expression we heard on several occasions in Ambassel and the neighbouring awrajas; cows have to be trained to do ploughing, a task which they do not normally perform.

According to peasant informants, the three most serious constraints to recovery are shortage of draught animals, of seeds, and the poor health of the peasants themselves. Any one of these factors will have serious adverse consequences on farm work and family income: less land will be brought under cultivation, the quality of farm work will suffer, and post-harvest losses will be high. Add to this the fact that pest infestation is generally greater in the early years of recovery, and you have some idea of how great is the burden that peasants have to bear during this period. But once again peasants attempt to overcome these difficulties through a variety of ways. Thanks to the tradition of mutual support, reciprocally beneficial cooperation, and the peasants' own ingenious farming practices, the Wollo peasantry has often succeeded in easing some of the burdens associated with post-famine recovery.

At the time of our field-work in the autumn of 1986, which was the first full year of recovery for peasants in Ambassel and Qallu awrajas, a great majority of peasants had planted sorghum, maize and horsebeans, despite the fact that crops like teff, wheat, lentils and peas fetched higher prices in the free market (see Annex 3). But many peasants favoured the former because they require less labour during planting and harvesting, and involve less cost in production. Moreover, the first two stalk crops can be stored without threshing and winnowing, or sold in the market in the same manner, and this reduces post-harvest losses. Furthermore, all three plants are stress-tolerant, and will mature in difficult conditions. Peasants who were hampered by the shortage of traditional power, seeds and family labour, attempted to overcome their difficulties by adopting production plans and crop mixes which involved less labour, less cost, and less loss.

About 17 per cent of our respondents in Ambassel reported that they had been unable to farm all or some of their holdings in the 1986 season due to the difficulties noted above. This is not a high figure considering the magnitude of the crisis in the area. Surprisingly enough, almost all those who said they had not farmed their land at all, or only part of it, were male; women household heads seem to have overcome post-famine obstacles far more successfully than their male counterparts.

Just as during famine it is the peasant community's own survival techniques that save it from large-scale disaster, so too during the post-famine period it is its own resourcefulness that enables it to recover and to ease the pains of recovery. While outside intervention has in the last two great famines played an important role in the transition from crisis to normalcy, recovery would have been a longer and more costlier process, involving greater stress on family and community without the peasant's ingenuity and effective use of individual as well as social resources. Insofar as recovery in production and farm work is concerned, the most important mutual support arrangements are the following: sharing draught animals, or renting them from others; share-cropping, or contracting labour for cultivation; leasing one's own land to others; and, individual or communal assistance.

About 63 per cent of the peasants we interviewed in Ambassel were able to cultivate their holdings, in whole or in part, because they acquired one or more draught animals from others through mutual support arrangements. One hundred and twenty-eight peasants, i.e. about 50 per cent of respondents, obtained such animals through rent. Table 28 provides a breakdown of the distribution of supportive arrangements for draught animals by woreda.

Of the peasants who managed to arrange access to draught animals, i.e. of the 160 peasants shown in the table above, 14 per cent were aged 39 and below, 24 per cent were in the 40 to 49 age bracket, 26 per cent between 50 and 59, and 36 per cent 60 and over. It seems that the difficulties posed by the shortage of farm oxen were evenly spread among all age groups, although, the under-39 group, which made up 17 per cent of our sample, appears to have been a little less hard hit, while the over-60 group, which was 33 per cent of our sample, a little more so.

Table 28. *Means of acquiring oxen (Ambassel 1986)*

Means	Ambassel	Worebabo	Tehule.	Total
Renting	24	49	55	128
Personal asst.	–	16	6	22
Communal asst.	4	3	–	7
Other	–	3	–	3
Total	28	71	61	160

We have discussed the various ways by which disadvantaged peasants in Wollo acquire draught animals, and there is no need to repeat the discussion here (see Chapter 5).

The great majority of those peasants who borrowed oxen agreed to pay for the service in labour, rather than in kind or in cash—to be precise, 86 per cent arranged to pay in labour, 12 per cent in kind, and the rest in other ways including cash. Again, women were underrepresented, as only 5 per cent of the peasants who made supportive arrangements to acquire draught animals were women. It appears that a higher proportion of women-headed households survived through the period of recovery by leasing their plots to others, or by entering into crop-sharing arrangements.

A few words should be said about the support measures that were provided in the form of individual and communal assistance. The first form of support was fairly widespread, benefitted about 14 per cent of the households in our sample, while the second was somewhat limited, reaching only 5 per cent of the beneficiaries. Individual assistance is exchanged usually among close relations, and often provided to the elderly, to widows in distressed situations, and to the disabled or the invalid. This is in reality outright charity which may be offered by one individual, or a group of two or more individuals. Communal support, on the other hand, is a group effort, and in a good number of cases, it was mobilized by the local PAs, either at the initiative of individual peasants, or of local extension agents. This form of assistance, which may be organized as *jiggae*—i.e. associated community labour—is of benefit to a large number of needy persons. In one PA in Worebabo woreda (which was not included in our sample) all the members were mobilized to work the land of those who were famine victims and short of draught animals and family labour.

It does not require an extraordinary effort for a peasant with two oxen to work his land and that of another family. Land renting, and setting up crop-sharing arrangements was not an uncommon practice in Wollo during the period under discussion. In a good number of cases it was the less drought-affected peasants in each area who were the most active in these arrangements. A peasant in this position may rent a poor or distressed peasant's plots for a flat fee to be paid in kind, in cash, or perhaps in livestock. The latter will not be involved in any aspect of the cultivation, as this is the responsibility of the renting peasant. In the crop-sharing arrangement, the benefit owed to the owner of the land for its use will depend on the harvest, and on the amount of labour and inputs he has contributed. Generally, it is not the shortage of draught animals that drives households to lease their land or to enter into share-cropping arrangements, but rather the inability to engage in labour due to old age, ill-health, or serious shortage of family labour.

Thanks to these mutual support and mutual benefit practices, far more land was brought under cultivation in Wollo in the early phase of post-famine recovery than would have been possible under the circumstances. Thanks again to these practices, seriously disadvantaged peasants, and families who had sustained heavy human livestock, and/or material losses were not abandoned to their fate.

How did peasants manage to have access to draught animals, and where did these animals come from? If we leave out communal assistance, the two major sources of oxen for renting to peasants were peasants from neighbouring communities and friends and relatives. The largest source of animals for rent were peasants in the first category, who supplied 60 per cent of the draught animals for rent to peasants in our sample, while in second place were friends and relatives with about 36 per cent. It is difficult to identify precisely who these sources of draught animals were, but a large proportion of them were: (a) peasants who had bought livestock from distressed peasants during the drought when prices were very low; and (b) peasants who were able to keep some or all of their farm animals through the crisis. In both categories must be included not only the peasants who were fortunate to have escaped the crisis but also those who where prosperous enough at the outset (in the relative sense) to weather the crisis and still retain some of their live assets. No outsiders were involved in these transactions, nor in any of the other mutual support practices under discussion. The whole operation was an inter-peasant affair, and if some benefitted more than others—and this certainly was the case—the "surplus" involved remained within the rural areas.

Famine has a levelling effect. We noted earlier that 63 per cent of the peasants in our sample were able to cultivate their land by renting oxen or with assistance from others, while 17 per cent of them had been unable to put part or all of their holdings into crops. From this and other evidence we have, we estimate that about half of our respondents had no oxen at the beginning of the postfamine period, and about 25 per cent had only one ox each. Among those who reported having worked their plots without support from others were peasants who used hand tools to cultivate their land because they were unable to acquire draught animals. The peasantry emerges from the famine thoroughly exhausted, and the general impoverishment that results affects everyone except those who were lucky enough to escape the disaster, and those whose survival measures were relatively more successful than their neighbours.

Famine also has a socially "de-stabilizing" effect, the net result of which is a restructuring of rural society, and a transfer of wealth from one sector to another. The losers are those who were forced to dispose of their assets during the crisis, or who lost them because of it. This group is made up of not just the poor peasantry but also peasants who were relatively more prosperous in the pre-crisis period. Indeed, the *new poor* stratum is made up of a cross-section of pre-famine rural society. Hunger will affect the poor and the less privileged more seriously than others, however, prolonged hunger, i.e. hunger that extends itself for more than twelve to fifteen months becomes less and less class-specific, and more and more general in its economic impact. The unique aspect of the food crisis in Wollo as well as the northeast was that it was: (a) a prolonged and intensified crisis which in some areas began in 1983 and did not abate until the end of 1985; and (b) for this very reason, it exhausted and ruined everyone that it affected, including a good number of peasants who were at the top of the rural social hierarchy earlier.

In contrast, the *new middle and upper* peasant stratum is a product of fortuitous circumstances (or the vagaries of nature), of economic activity during the famine, and strong economic position or kin network in the pre-crisis period. Specifically the new stratum consists of: (a) Peasants who escaped the crisis either because their communities were not hit by drought, or because they had a common natural resource (like a river, lake, spring, etc.) which they employed to fight against the effects of drought. Among these peasants may be those who were men of modest means before the crisis. (b) Peasants who took advantage of the behaviour of the market during the crisis and engaged in vigorous commercial activity. Peasants in this category may be the same peasants as in the first category; what they did was buy one product in one market and sell it for profit at another, trade in high price products, especially grain, or buy livestock cheaply and keep them out of the famine zone for the duration of the crisis. (c) Peasants who had enough assets and came out of the crisis with limited damage. Such peasants will have employed a more successful survival strategy than others owing partly to their strong economic position. And finally, (d) peasants who had a large number of kin and relations, particularly outside the famine zone, who provided assistance of various sorts. These peasants may have been well-to-do *or* poor before the onset of the crisis.

Here again, the new well-to-do stratum comes from a cross-section of rural society and it is peasants in this social group who play an active role in post-famine recovery. They lease draught animals to others, they rent land from those who are not in a position to farm, or enter into crop-sharing arrangements with them, etc. The stronger in the group engage in livestock trading either locally or in distant area markets, and also in money-lending activities, as we shall see below.

It was noted above that there is a transfer of wealth from one party to another, but this now takes a new form. The subject may be looked at in two ways: transfer of wealth vertically from the new poor social group to the new "prosperous" social group, and horizontally or spatially, from the post-famine zones to the famine-free zones. In the first instance, the well-to-do gain greater benefits when they become involved in the various mutually supportive practices discussed above, although the service they render in the process is also indispensible to the poor peasantry. "Wealth" here may also include labour, which for example, the poor peasant renders to the farm animal owner from whom he rents oxen. For the most part, however, "wealth" consists of "surplus" products, largely food grain and other agricultural goods.

In the second case, resources flow from the food deficit areas or regions to the food surplus areas. These resources consist of cash, live assets, and high-value personal belongings. The real magnitude is difficult to estimate but the net value of resources that flowed into south-western Wollo from mainly central and south-eastern Wollo during the famine years and after was quite considerable. Similarly, there was a net outflow of resources from the famine zones of north and north-west

Wollo to the food surplus zones of Gondar and Gojjam provinces, though on a smaller-scale. The main beneficiaries in both cases were peasants, and rural traders in the recipient regions. These individuals had two advantages over their unfortunate neighbours: surplus food, which they sold to distressed peasants at high prices, and cash (mainly the proceeds from the marketing of grain) with which they bought livestock from these same peasants at very low prices. The famine stricken regions will continue to suffer loss of resources as peasants here will have to purchase livestock from peasants and rural traders in the distress-free zones.

The post-famine period is also a period of restocking as peasants are eager to make up their losses, and to improve their level of income and well-being. But this is a difficult task since most peasants are destitute, and since there are very few income earning opportunities outside agriculture, and hardly any disposable assets in the communities. Peasants who bought farm oxen after the famine did so through complicated installment-payment arrangements which allow them to pay for the animals over a period of time. Others who purchased other livestock acquired the cash by selling some of the relief rations they received.

Far more peasants would have taken out loans than actually had in 1986 if they had more and better access to credit services. Local money-lenders were not fully active in this period, and neither the Service Cooperatives nor Peasant Associations were in a position to provide loans to their members; the same was true of government extension agencies.

About a quarter of the peasants in our sample said they had borrowed money in the first year of recovery, i.e. 1985/86. Of these, 75 per cent had borrowed from relatives and friends, 17 per cent from urban-based merchants, and 8 per cent from peasant money-lenders. The term "relatives and friends" should be taken to include peasants themselves, and persons living outside the rural communities, but there is reason to believe that the majority were among the former group. Taking this into account, one can say that in the majority of cases, the sources of credit were the peasants themselves.

Peasants borrowed money for a variety of reasons, but interestingly enough very few used borrowed money to purchase draught animals. This might be because farm oxen were priced high and neither the borrower nor the lender could afford to be involved in this sort of heavy transaction. As Table 29 indicates, many peasants took out loans to

Table 29. *Reasons for peasant borrowing (Ambassel 1986)*

Reasons	Borrowers	
	No.	%
Buy seed	30	34
Buy food	27	31
Pay tax	21	24
Buy oxen	7	8
Other	3	3
Total	88	100

Note: Peasants had borrowed more than once.

cover necessary expenses, rather than to invest or restock. This may be partly due to the impact of relief activities, and partly to the levelling off effect of the food crisis noted earlier.

About one-third of the borrowing by peasants went to purchase seeds for planting, and, surprising as it may seem, 31 per cent was for buying food. It should be noted that peasants in most of the famine-stricken areas of Wollo and the northeast contined to receive relief food aid at least until the end of 1986, and one would think that they would not go into debt to buy food. The explanation for this seeming anomaly is that peasants sold their relief rations and used the proceeds to purchase livestock. In most instances the food distributed by relief agencies was not the staple food of the peasantry anyway, and peasants often marketed it and bought what they preferred in the market. Peasants unable to maintain their families until harvest must then have borrowed money to purchase food.

The other major outlay for which peasants went into debt was the government tax. We shall return to this point later. In conclusion, it may be said that the struggle for recovery was a difficult one, and the burden the peasantry had to shoulder to emerge from the depths of despair and destitution was heavy and debilitating. About 63 per cent of the peasants in our sample were able to farm their land by renting draught animals or with assistance from others. Some 25 per cent of these same peasants were in debt, this despite the fact that emergency relief was still distributed. About half of our respondents were without draught animals even after a few thousand oxen were distributed free to peasants in the area by several NGOs. Obviously the poor condition of the peasantry will continue to have adverse consequences for a long time to come, and real, full-scale recovery may not be achieved in the immediate future.

Emergency Operations and NGOs

That very little has been said about emergency relief, and relief-related rehabilitation activities should not be taken to mean that these activities have not played an important role during the crisis as well as in the initial period of recovery. The reason is that the subject is not part of our study, and that it deserves to be treated extensively and preferably by itself.

The story of the colossal relief operation that was undertaken by UN missions and donor agencies, and the critical role played by volunteer relief organizations, or NGOs, has yet to be told in full. Short and fragmentary accounts of what may have been the largest relief effort in Africa in this century have appeared in the last few years but these provide only a partial picture. Whatever may be said against it, this great human undertaking will remain a success story, and a fine example of human co-operation.

When relief operations began in earnest in October 1984, RRC was supported by about twenty to twenty-five NGOs in the distribution of food and other relief supplies to an estimated needy population of over five million. By the end of the year and early in 1985, there were nearly sixty-five NGOs and more than half a dozen UN agencies engaged in relief work throughout the country, with the majority deployed in the northeast and in Eritrea. At the beginning of the programme, the division of responsibility agreed upon between RRC and the NGOs was that the former would be in charge of general dry ration distribution while the latter would be assigned to special feeding and medical care, but as the burden of the operations became heavier NGOs became increasingly involved in all aspects of relief work. By the latter part of 1985, NGOs were responsible for distributing 65 to 70 per cent of all relief supplies in the country. This was quite impressive considering the fact that the total number of expatriate and local staff working for the volunteer organizations at no time exceeded two thousand, while RRC on the other hand had an established nation-wide field network and operated with a staff numbering over fifteen thousand. RRC's relief operations faltered during 1985 because it was burdened by its own bureaucracy, by the severe shortage of trucking capacity, and because it was pressed into giving priority to resettlement activities which were launched in December 1984.

Not all volunteer organizations which flocked to Ethiopia had noble objectives nor did all of them set examples for high standards of efficiency and management. A good number of NGOs and the volunteers working for them were in on the act for the adventure, the publicity it would provide, and for political or religious inspired motivations. Some of the NGOs had very little previous experience in relief work while others had extensive work experience both in Ethiopia and other Third World countries. Kurt Jansson, whose responsibility it was to coordinate the work of the NGOs speaks highly of some and critically of others. He singles out, for instance, CARE, the International Red Cross, OXFAM, and Save the Children Fund to praise their competent leadership and management. His strongest criticism goes to the French-based, right wing relief agency Medecins sans Frontier or MSF which was involved in health work in Korem and Qobbo. This organization was made up of inexperienced youngsters straight out of medical school, and it was expelled by the government in December 1985, the first and only NGO to be thrown out of the country during the famine. Jansson points out that some of the staff of MSF "were highly excitable, reacting emotionally to any events that their Western stands did not think appropriate". They fed the French press, he says, with stories which were either patently false or highly exaggerated. "The MSF was criticised by many NGOs and some of them felt that the organization actually wanted to be expelled" (ibid.: 25).

Beginning in the first quarter of 1985, tension began to build up between the NGOs working in Wollo on the one hand and RRC and the local authorities on the other. The Wollo administrative authorities in particular were far more concerned with collecting information on the relief operations than in the operations themselves, and insisted that NGOs provide the necessary data sometimes on a monthly basis, something which the agencies were not prepared to do. Some of the documents prepared by NEERNDRC single out for criticism several NGOs, including World Vision, German Emergency Doctors, and Missions of Charity for failing to "cooperate" with the authorities. NGOs also served as scapegoats for unexpected mini-disasters that occurred during the famine. The cholera epidemic which killed large numbers of famine victims in Wollo in the first half of 1985 caught the Wollo authorities unprepared, but they blamed the poor handling of "inexperienced foreign medical staff" for the high rate of deaths, particularly in Korem where some 18 per cent of the cholera patients are said to have died (NEERNDRC 1985 A: 17). Michael Pelly, a medical officer who worked briefly for SCF in Korem and who was present at the time of the epidemic says however that the death rate was kept low thanks to the energetic efforts of the NGO medical teams working there.

The tension between the NGOs and local authorities was further exacerbated by the sheer bureaucracy of the latter, and by the ongoing conflict and rivalry between RRC and the Ministry of Agriculture (MOA) at the provincial and local levels. NGOs are registered with RRC and assigned by it to specific areas for relief and rehabilitation work. But MOA argues that PAs, with which NGOs have to work, are under its jurisdiction and any rural programmes involving them have to be approved by it. The attempt to negotiate one's way through the bureaucracy and the existing rivalry was a frustrating experience for many (see, eg., JVC's monthly reports for 1986).

The work of NGOs and UN missions was further hampered by a variety of other difficulties. The government collected revenue on relief goods, including food aid, and port charges on all relief cargo were very high. It took a great deal of lobbying by UN missions, and some persistent protests by NGOs before the government grudgingly agreed to reduce port charges by 10 to 17 per cent in November 1985 (see *NGO Newsletter*, October 1985). But the most serious threat to relief operations was posed by anti-government forces operating in the north of the country, and old style banditry in the rugged areas of Wollo and Gondar provinces. In 1983 and 1984, anti-government elements raided Korem, Awarae village in Raya Qobbo awraja, the Qobbo-Alamata agricultural scheme, and Tis Abalima and Jari in Ambassel awraja (all in Wollo province). Shelter camps were attacked, relief supplies were looted, agricultural schemes and infrastructure destroyed, and some foreign relief workers kidnapped. In both these years many NGOs had to leave north Wollo because of the security breakdown there.

In 1986, and in the last quarter of 1985, relief agencies were distributing not only food but draught animals, hand tools, seeds and fertilizers. About 2,800 oxen were distributed in Ambassel and Qallu awrajas in the period up to October 1986, but this probably reached less than 1 per cent of the needy population of the area. A few hundred oxen were also distributed in Ambassel by government authorities but the beneficiaries were all producer's cooperatives and not individual peasants

(NEERNDRC 1985 A: 59); a smaller number of oxen was similarly distributed in Qallu awraja by the same authorities (NEERNDRC 1985 QA: 75).

The importance of emergency relief was that it played a positive role in post-famine recovery. It enabled peasants to invest in livestock (mainly sheep, goats or calves), and provided to a limited extent, an alternative source of disposable income. There were a number of extension agents at the time who felt relief activities should be suspended on the grounds that the peasant was becoming dependent on external aid and was not giving his plots sufficient attention. This argument was of course spurious, and there was no evidence that peasants had willfully neglected their farms, or that they intended to live off relief aid for the rest of their lives. What relief did was to provide that small margin of income which the people used to invest on their farms and to improve their production.

All through the previous chapter as well as this one we have emphasized that survival and recovery (or partial recovery if you will) was achieved predominantly by the peasant's own efforts, and bringing into full play his own moral and social resources. This is not to deny the important role of external intervention and assistance which for the most part consisted of relief food aid donated by Western governments and UN agencies (see Annex 6). The point is that these were supplementary support measures, and did not constitute the central element of indigenous survival or recovery strategies.

This brings us to the role local authorities did not but should have played to ease the burden of the peasantry during postfamine recovery, and to accelerate the return of rural Wollo to normalcy. The first, and the most common measure governments adopt when famine strikes is to relieve the peasantry of the burden of taxes. This is a well-known short-term remedy and has often been carried out in many countries. It is not a very effective measure in itself, but it relieves the peasant of some of his difficulties, and it conveys to him the message that government and society are aware of his conditions and conerned about his problems. As was observed above, the peasantry in Wollo and the northeast was not provided tax relief, nor was it relieved of the famine levy, or of other dues it normally pays locally. Moreover, the peasantry continued to be burdened with corvée labour, as discussed in Chapter 5 of this study, and was still subject to mobilization by local extension agents and peasant associations for the purpose of being put to work on community projects. This, despite the fact that the people of the region had just emerged from a tragic and exhausting experience, and were barely capable of working their own fields.

The other policy issue of concern has to do with the attitude of local officialdom to the rural market and market system. In some parts of the northeast, and in Wollo as well, there is a growing conviction among state bureaucrats, particularly those in planning and administrative areas, that the peasant is using the market for wasteful purposes, and that time and effort which should have been profitably used in farm work is wasted in attending a variety of markets. As a result, plans are being worked out (in some areas they are already being implemented) to reduce considerably the number of markets held in a locality, and to force peasants to attend only one market a week, and this on a specific day designated by local officials. If these plans are fully implemented, local officialdom will succeed in seriously crippling, if not completely destroying, the rural market system, one of the most critical "weapons" of the peasantry in its fight against death and privation.

There is no need to repeat it here, but we believe we have shown with sufficient clarity that the rural market system is an important factor in the peasant's resource management strategies, and tampering with it will have adverse consequences on the economy of peasant households. The arguments of local bureaucrats regarding the wasteful nature of the present market system is of course without foundation, so too their plans, which, based as they are on these unfounded arguments, will only lead to the destabilization of the peasantry and the further depression of the rural economy (see Chapter 5 for a discussion of the market system).

IV CONCLUSION: BEYOND SURVIVAL

10. Neither Feast Nor Famine

We shall bring this study to a close with a short discussion of the potentials of indigenous survival strategies in combatting future famines, and the policy options available to the country if the goal of a hunger-free society is to be achieved in the not too distant future. It is not our intention to offer a blue-print for national salvation, but at the same time we believe there are clear and concrete lessons to be learned from the tragic disasters that have afflicted this country in the past, lessons that point to definite policy measures, and sustainable defensive strategies. Informed opinion here, at least a substantial part of it, is agreed that the scourge of famine will continue to haunt this country for many years to come, and it is incumbent upon government and concerned authorities to make the fight against hunger their highest priority. It may be stating the obvious but it is worth pointing out that what the peasants of this country are facing is a crisis of livelihood which, left unresolved, will come to threaten the very survival of rural society if not society at large.

But the most determined effort to solve the problem will have limited results unless the experiences and lessons of the past are carefully examined, and until the peasant himself becomes the key element in a new, flexible and dynamic anti-famine strategy. We believe the experiences of the peasant can and must be incorporated in a coherent and revitalized food system policy, and a policy of sustainable growth.

We may preface the discussion with a brief comment on the theoretical explanations of famine and food shortages provided in the current literature. The aim is to place the Ethiopian experience in proper perspective. This is not the place for a fullscale exegesis of the major works on the subject, and a short review of the chief tendencies presently in vogue will be sufficient for our purposes. There are, broadly speaking, four main currents of opinion. Each has been argued and defended more or less extensively both in public and within academic circles.

As a result of the several major famines that have occurred in the last three decades, and owing also to the even more numerous localized food shortages as well as cases of serious malnutrition endemic in the Third World, the issue of food and food distribution has aroused a great deal of interest and has become a subject of public debate in many countries and within the UN and its agencies. The first viewpoint, which may be labelled populist or left liberal, is, to a certain extent, a product of this heightened public interest on the ethics and politics of the world food order (see Friedmann for a historical analysis of the world food order).

The proponents of this viewpoint stress that there is hunger and privation in Third World countries because some of the countries of the West have more food than they know what to do with. According to these writers (George, Lappe and Collins, Dinham and Hines), the increasing incidence of hunger in the world is largely caused by powerful global forces—major food producers, multinational agri-business, etc.—which have a stranglehold on both the supply of food and the movement of world food prices. This direct or indirect domination of the world's food resources (a) induces poor countries to neglect their own agriculture and become dependent on food imports and food aid; (b) encourages rapid population growth and urbanization in food deficit countries, hence greater demands for food; and (c) encourages greater attention to be paid to the production of cash crops as opposed to locally consumable food crops.

In the second place, there are those who argue that the cause of the food crisis in the Third World is tied to the development of capitalism in the countries concerned, and to their subordination and incorporation into the world capitalist system (see special issue No. 15/16, 33 of *Review of African Political Economy*, Watts). This is the most dominant outlook on famine within left scholarship, not, we regret to say, because the arguments are convincing, nor because the empirical evidence supports these propositions, except in a few obvious cases, but largely because it has become conventional to attribute all the failings of Third World countries to external forces, and to absolve the internal and intrinsic social forces and political and economic structures from blame or responsibility. Indeed, on occasion, careful investigation, and reasoned and informed analysis is sacrificed in favour of the rhetoric of external dependency.

A slightly different version of this school of thought attempts to look at famine as a by-product of the relation between the peasant and the capitalist modes of production, or the articulation of one and the other (Wisner 1977). The blockage of one mode of production by the other, or dependence of the weaker on the stronger is said to lead to food crisis.[1] We shall see later that of the numerous countries that have experienced hunger only a few fit this model.

The third current of opinion attributes famine to poverty, inequality, failure of entitlement and social injustice, (Dumont and Cohen, Sen, Spitz). The proponents of this view emphasize that famine is primarily the outcome not of food shortages but rather of poverty, and lack of purchasing power—in a word underdevelopment, with the accent, at least according to Sen, on deprivation of access to the sources of livelihood. There is some debate within this school as to whether food crisis occurs as a matter of course or is preceded by a sharp decline in the per-capita availability of food but Sen, against whom this argument is made, stresses that this is not the case while his opponents remain unconvinced (Sen 1986 and 1987; Alamqir 1980, Bowbrick 1986, 1987). The other school of thought emphasizes that underdevelopment breeds its own form of social structure and property relations which often deny the potential victims of famine access to property and sources of employment. Underdeveloped countries are not only poor but also dominated by grasping elites, often non-peasant in origin, who control vast agricultural resources. The problem with this line of thought is not that it is invalid but many of its arguments have been raised before, and the substance of many of its conclusions are hardly new. As far back as the early 1930s R.H. Tawey masterfully showed the intimate linkage between underdevelopment and mass starvation in his classic study of rural China.

The fourth current of opinion, which is mostly of historical interest, associates famine with colonialism and the destructive effects of colonial policy on African or Asian agriculture. While most proponents of this view have been content to attribute the pre-independent famines in the Third World to the ruinous and exploitive policies of colonialism some

[1] Lars Bondestam whose sophistry concerning capitalist development in Ethiopia we exposed in an earlier work (Dessalegn 1986 [A]) now offers another in connection with food. The solution to the food problem in Third World countries, he says, is socialism adapted to the conditions of each country. He may not be informed of it but of the ten or so socialist countries in Africa, only one managed to escape severe famine in 1984. See below.

have extended the argument, and still see the hand of colonialism in the food crises of postcolonial Africa or Asia (e.g. Rotberg 1983). The colonial legacy is believed to continue to unfold its insidious effects long after colonialism itself has disappeared.

It is difficult to place Mesfin Wolde Miriam (1984) in any of these four currents of opinion although he is in many ways closer to those who equate food crisis with destitution and underdevelopment. The main problem with this work, however, is not that it does not fit into any school of thought but rather that it does not really accomplish what it sets out to do, namely provide a new theory of famine causation. The central proposition is couched in Manichean terms, with the peasant, the victim of exploitation on the one hand, and what are called the political, social and economic forces, the perpetrators of injustice, on the other, tied up in a relationship in which the former always loses everything, including the ability to feed himself. The forces arrayed against the peasant include everybody except the poor peasant, the victim of expropriation. As we have tried to show, however, there is as much co-operation and mutual support within rural society as hostility and surplus expropriation, and that, for example, the peasant on the one hand and the rural trader on the other may be antagonists today but "allies" tomorrow. There are as much individual and inter-communal supportive and reciprocal relationships as antagonistic ones. Moreover, famine does not always lead away from co-operation and association to individualism, as has been claimed by some (Spitz).

Looking at famine in a historical perspective, and within a macro-regional context, we find that the problem of causation is far more complicated than is indicated by the theoretical debate noted above, and that famine has occurred in a large variety of historical, political, economic and ecological contexts. The following is a summary presentation of this:

1. Famine has occurred in the same country, region or geographical zone before, during and after colonialism. The evidence now available shows, for example, that the earliest recorded famine occurred in West Africa in the 15th century. This period witnessed the rise of settled kingdoms, and the expansion of long-distance trade in the region. One of the most serious famines in the Sahel, known as the Seven Years' famine, occurred in the first decade of the 18th century, again before the appearance of the European or colonialism in the area (Schove). Simi-

larly in India, famine has been shown to have taken place in the 10th century A.D., and one of the worst famines India has suffered is said to have occurred in 1343–45 (Alamgir, Manetsch). As far as Ethiopia is concerned, the history of famine goes back to the 9th century A.D., and by the 15th and 16th centuries, famines and epidemics had become regular occurrences (Pankhurst 1985).

Enough has been written about the famines that took place in Asia and Africa during the colonial period, and there is no need do dwell on the subject here (Alamgir, McAlpin, Rotberg, Sen 1981 B, Watts, Vaughan). In the post-colonial period, the serious food crisis in Bihar state in India in 1967, and the devastating Bangladesh famine of 1974 are worthy of note in the case of south Asia. Post-colonial Africa has experienced at least one famine every decade since the 1960s, the famines of the early 70s and 80s being the most damaging.

2. Famine has occurred in countries which have had no colonial experience. The history of famine in China goes back to the middle ages, if not earlier, and the worst famines which claimed millions of lives occurred in the middle of the 19th century (Tawney). The case of Russia is even worse, at least as far as the frequency of occurrences are concerned. There were seven major famines between the 12th and 16th centuries, four in the 18th, and thirteen in the 19th century. The last famine before the Bolshevik revolution occurred in 1911 (Fisher: Ch. XXI). A third country without a colonial past but with numerous famine experiences is of course, Ethiopia.

3. Famine has occurred in countries which were not seriously integrated into the world capitalist system. The example here may be Ethiopia (the famine of 1974), and Bangladesh (1974). While it may be argued that Russia before the revolution was experiencing the growth of capitalism (though by no means fully capitalist), it would be unrealistic to include rural Russia as having been within the global capitalist orbit.

4. Famine has occurred in countries which for many years had pursued socialist policies. The first famine revolutionary Russia experienced was the one that occurred in 1919/1921 (see Edmondson). Some may attribute this to the civil war and the disturbances that accompanied the First World War, in which Russia participated, and the Bolshevik seizure of power. But there was a second, though less well-publicized famine which took place in 1933, i.e. more than a decade after socialism had consolidated itself (Nove: 179). In both cases the worst hit areas were the traditional grain surplus regions of the country.

Recent evidence has revealed that China was struck by a crippling famine between 1958 and 1961. If the estimate of the death toll now considered is even half true—i.e., more than 6 million persons—this famine becomes the greatest peace time disaster to have occurred in this century (see Ashton *et al*). On the other hand, almost all the socialist countries in Africa, including Ethiopia, Tanzania, and Mozambique, suffered seriously due to the famine of 1983/85.

This brief exposition hardly does justice to the debate in the literature, however it may indicate the complexity of the problem, and the inadequacy of the single factor approach to famine analysis. The common factor in all the cases cited above may be relative rural backwardness, but while all famines (or almost all of them) have occurred in underdeveloped conditions, not all underdeveloped countries have experienced famine. The Ethiopian experience further complicates the issue: famine has occurred throughout this country's history, under a variety of political, economic and social systems (traditional and modern), and in times of war as well as in times of peace (Pankhurst 1985, Charles Wood).

We have tried to show in this study that Wollo (the northeast as a whole) will not become a land of plenty in the foreseeable future, but at the same time, we are not convinced that it is condemned to remain a land of scarcity forever. If there is no feast, then there will be famine: such is not, and should not be the fate of the region. The peasantry of the region is capable of self-sufficiency in food—self-sufficiency measured in relative terms—and a realistic programme of rural development should aim at enabling the people to reach and maintain this goal. We are speaking, in other words, of *sustainable* survival strategies, and *sustainable* development. This concept, sustainable development, may soon become a cliché—thanks to bureaucrats in the specialized agencies of the UN—but when it was originally employed, it referred to the need to harmonize what is necessary with what is possible (Redclift 1986). Specifically, sustainable strategies are those which will enable a population to maintain a tolerable level of productivity and income in the face of stressful social conditions and/or environmental crisis.

There is another conservation problem that has not often been considered, that is food conservation as opposed, or rather as a complement to, soil conservation. While crop losses due to poor crop storage may not be intolerably high in general, there is sufficient loss to hurt the producer (see the debate on the subject in IDS Bulletin, Vol. 13, No. 3, 1982); this loss could easily be prevented with better or improved grain storage, and better techniques of storing at the household level. A far more interesting proposition could be better or improved storage at the communal level through the agency of the Rural Service Cooperative—we shall return to this subject later. A programme of food conservation should include, in brief, a) a reduction of pre-harvest losses, particularly soon after drought when pest infestation is very high, and b) a reduction of post-harvest losses through better individual or community storage.

We have shown earlier how the *rural market system* is a critical element in peasant survival strategies, and how it also plays an important role in post-famine recovery. It is essential, therefore, that the free flow of people, goods and ideas within regions, zones or communities must not be hindered for whatever reason except those having to do with security. Over and above the economic advantages, inter-regional and inter-community trade will help break cultural barriers, ethnic and historical prejudices, and will bring people together in closer community.

Related to this is the issue of *agricultural prices and government requisitioning*. We will not spend time discussing these two issues as they are outside the scope of this study. Our proposal on the subject is simple and direct: policy-makers should be persuaded as soon as possible to scrap completely the food grain price system now in force in the countryside as well as the system of forced grain requisitioning. These two policies have caused, in our opinion, incalculable damage to the peasant economy and to smallholder agricultural development. Both should be replaced by a free market and what we call an *incentive price* system (see below).

A third support measure which also is not new, and fairly commonly discussed in the literature is the introduction and popularization of *high-level drought tolerant seeds*, and *famine crops*, i.e. plants that grow in difficult environmental conditions, and which require less labour and less cost. It is true that peasants will not readily adopt new plants and new sources of food, especially if these call for new or different working practices, and a different agricultural calendar. New crops that are considered for adoption, therefore, in the various agro-ecologies of northeast Ethiopia should have characteristics that fit in well with the established cultivation practices of the peasantry and the normal agricultural calendar of the various zones. Additionally, however, the price incentive could be used to induce peasants to grow the new crops on a regular basis, and this could work in two ways. Peasants could either be offered the new seeds or plants at very low prices so as to encourage them to try the crops, or an authorized agency could purchase the new crops from the peasantry at reasonably high prices. Provisions could also be made to compensate peasants whose experiments with the new plants fail.

This fourth proposal will involve major policy revisions, as well as a new and meaningful outlook on poverty and its alleviation on the part of government and international donor agencies working in Ethiopia. It was argued in Part II that among the chief obstacles to peasant security and growth were insufficiency of holdings, low disposable income, low consumption—in a word poverty and destitution. One way to alleviate this acute poverty in rural Ethiopia is to designate a poverty line and to allow an individual household living below this line to be relieved of all state taxes and local dues. Since owing to the poor state of information on income and consumption, it will be difficult to determine this line on the basis of liquid assets or annual income, the government should use size of land holdings as an alternative criteria. A minimum holding sufficient for a family, for example, will vary from one ecology or agro-system to another, but the following could serve as a working model: a minimum holding sufficient for a family of five in a high density cereal zone: 0.5 ha; in a high density ensat and/or cash crop zone: 0.25 ha; in a low density cereal zone: 1 ha. In Wollo, which falls in the first category (so too most areas in the northeast) any household which possesses land measuring half a hectare and below should be declared as living below the poverty line, and exempted from all taxes. The government might be reluctant to introduce such a policy as it will mean a loss of revenue, but international donor agencies could offer it incentives and support measures that it may find appealing.

The foregoing are general recommendations that are meant to complement the support measures provided below. These measures are based on the specific agricultural practices and famine experiences of Wollo and the northeast which were discussed earlier in this study, and which will not be reproduced here. It is assumed that the reader is familiar with the issues and problems involved.

Disaster designation and early warning

The most important support activity that the peasantry will benefit from greatly is in the area of disaster anticipation and disaster preparation. There are several immediate and emergency measures that can be taken in the event of a crisis, or impending crisis. The first step is to introduce the practice of designating certain areas as *food emergency zones*, on the basis of prior determination. A government, para-government or locally-based international body could be authorized to assess crisis prone areas, and once such areas are reported to be suffering environmental stress of sufficient magnitude they should be designated food emergency zones. Such zones should, by law, be beneficiaries of a variety of assistance programmes, including tax relief. A few, select areas or regions, such as the lowland parts of Tigrai province, and eastern Wollo (specially, eastern Qallu, and most of Aussa awrajas) should be designated *permanent food emergency zones*, and offered the same benefits on a long-term basis.

A more effective, though less dramatic support is to strengthen the *early warning capabilities* of the peasantry particularly in those regions with a history of famine and environmental instability. It would be too much to expect RRC to provide reliable early warning information to the peasantry on a regular basis, but help could be provided to the people in acquiring information of this sort by themselves. This could be done in a variety of ways. We have seen that peasants put great store by the behaviour of the rains in the spring season, the reason being that the condition of the rains in the season is an indicator of things to come in the autumn season. With a little bit of training and some material support peasants can be enabled to keep more accurate records of environmental behaviour in their localities, and with this kind of record more meaningful assessments vould be made.

The second way to strengthen the peasant's early warning capabilities is a little more involved. We suggest as a first step that RRC be equipped at the provincial level with early warning assessment and monitoring capabilities which it does not have now. This could be done on a gradual basis, first equipping the famine prone provinces such as Wollo and Tigrai, then later the others. The programme should not be an elaborate one—the more modest the better. A few trained individuals could be attached to the RRC office in the provincial centre, whose job would consist of compiling early warning information in the province, using all available techniques including ground surveillance. It is pointless to have an early warning capability located at the head office in Addis Ababa, far from the centre of disaster activities. The provincial office will of course be expected to report to the head office on a regular basis.

This new early warning system will be effective if it involves the peasantry itself. We suggest that an early warning unit will be formed at the level of the Service Cooperative (SC). An SC consists of a group of anywhere between four and six Peasant Associations (PAs), and is empowered to collect and store grain, and to provide marketing, consumer goods, and other services to the peasantry. A woreda may have three to five SCs, depending on its size and number of PAs. One or two peasants, attached to the office of the SCs should be given appropriate training to enable them to record all pertinent information necessary for disaster assessment in their SC areas, and to report this information directly to the provincial office of RRC periodically. To avoid bureaucratic delays, the SC early warning unit should bypass the woreda and the awraja and connect with the provincial RRC which should in turn deal directly with the SC. Any time it is deemed necessary, the provincial RRC early warning and surveillance team could visit the SC considered to be threatened with disaster to make a field assessment. To make sure that the SC early warning team is providing accurate information the provincial RRC office should carry out spot checks in the SC periodically.

The SC team should also provide (and periodically update) the following kinds of information to the provincial RRC office. This is to enable RRC, and through it volunteer organizations to quickly and effectively channel emergency assistance to the areas concerned. This information should be compiled by RRC and made accessible to all. The important elements of the information needed are: (a) the characteristics of the population in the SC, and its geographical distribution; (b) the main crops grown in the area, their varieties and special characteristics if any; the food habits of the people, including preferences imposed by religious faith; (c) grazing and rangeland conditions; sources of water for agriculture as well as for human consumption; and (d) transportation, including rural road networks; facilities for stocking food and other emergency supplies; markets and the market network in the area.

Once the early warning capabilities of the peasantry have been strengthened, the SC could be assisted in establishing an *emergency seed bank*. This bank should stock up: a) drought resistant seeds, and/or b) seeds that RRC or other concerned authorities wish to encourage peasants to grow in a particular area. When the spring or autumn rains behave badly peasants should be allowed access to the seeds in this bank either for free or at nominal prices. Peasants should also have access to it during the phase of post-famine recovery. RRC or aid agencies could administer such banks as part of a *Strategic Seed Reserve* system.

Related to this is the question of *rural credit*. That a peasantry often suffering from disaster and privation should have access to cheap and easily available credit will not be an issue of dispute, but there may be differences of opinion over the question whether such a facility should be incorporated within the SC or the individual PAs. If rural credit is handled by the individual PAs, the advantage to the peasant is easy access, less bureaucracy and hence less delay; the advantages to the credit suppliers are less "mismanagement" (i.e. less loss through corruption), and easier administration. On the other hand, the SCs have a good deal of experience handling (and also mishandling) money or liquid assets, and those involved in promoting the idea of rural credit may want to take advantage of that.

Let us now move on to the issue of what may be called *strategic food reserve*. The Old Testament story of how Joseph saved the people of Egypt from starvation is worth pondering about *even* today. What he actually did was to build up a strategic food reserve during the years when the harvest was good, and dispense this food to the hungry during the years of death. We suggest a similar food reserve capability on two levels: at the level of the Service Cooperative, and at the national level through the agency of AMC. Both reserve systems will work efficiently if the policy changes regarding pricing and requisitioning noted above are put into effect.

The SC is an ideal institution for a locally-based system of food reserves, which should work as follows. The SC will actively buy food at free market prices from the peasantry as a matter of policy. This food is stored, and sold to the peasantry either at subsidized prices or on credit during hard times. The peasantry should be encouraged—especially through favourable prices—to sell to the SC routinely, and also to buy from it whenever it wishes. The SC should be assisted in building small but efficient storage facilities for stocking food grain, and given the chance to play the role played by the average grain merchant with whom it should actively compete. Food deficit areas will have difficulty establishing the system initially, but once it is off the ground the system should run smoothly. NGOs and aid agency could help launch the system by assisting SCs to build efficient and effective storage capabilities.

This system has the added advantage of enabling the peasantry to cut down its post-harvest losses. There is no reason, once the local food reserve system is in operation, why peasants should store their disposable, "surplus" grain individually. They could sell this grain, and even their own to-be-consumed grain, to the SC and buy it back if they wish and when they wish at current free market prices. Thus a community could own one "collective" (and efficient) store instead of many individual, and traditional ones.

The second type of strategic food reserve system should be set up either by using the present structure and capabilities of the AMC, or preferably by transforming the AMC itself from what it is now to an agency responsible for food reserves on a national level.

What is amazing about AMC is not that it is an inflated and unefficient bureaucracy, but that it has been engaged ever since its inception in siphoning "surplus" food out of the famine zones to feed the people of the cities. This has happened even when men and women were dying of starvation in these zones. Wollo, for example, is the fifth largest provider of surplus food to AMC, and since 1980 it has delivered to this agency 595,000 quintals (59,500 metric tons) of food. This food, which was taken out of the province, and out of the starving hands of the peasantry, was 40 per cent more than the total emergency food aid that was delivered to Wollo (about 36,600 metric tons) and distributed to 1.2 million peasants in the Ethiopian year 1977 (1983/84) (NEERNDRC 1985 A: 10).

Most of Shoa, Wollo and a good portion of Gondar province were in distress beginning from 1981, and yet the people were forced to give up what food they had to AMC every single year starting from 1979. We believe this agency could play a more constructive role if it was actively involved in the food reserve system noted above.

AMC should establish a network of strategically located storage facilities. It should buy food from the peasantry at free market prices for storage in these facilities. How much it should buy in a given year will depend on the state of food production in the country in that year. In difficult years AMC should offer peasants *incentive* prices for their grain, i.e. prices slightly above the market price. This stored food should then

Table 30. *AMC purchase of food grain from the five main provinces 1979/80–1985/86 (in thousand quintals)*

Years	Shoa	Gojjam	Arssi	Gondar	Wollo
1979/80	856.8	368.8	165.7	75.8	88.6
1980/81	682.2	853.7	282.1	163.6	98.6
1981/82	560.9	811.6	596.5	263.2	171.5
1982/83	1030.3	1171.6	828.3	315.9	154.7
1983/84	683.3	1015.6	500.0	190.9	24.4
1984/85	360.0	428.5	169.4	79.6	4.6
1985/86	887.6	814.0	537.9	162.2	52.2

Source: Fassil G. Kiros and Alemayehu; AMC June 1985.

be sold to the peasantry, preferably at subsidized prices, in times of food shortage.

Both this and the local system discussed above have the advantage of stabilizing prices during crisis and of discouraging speculators and grain merchants from driving up prices unreasonably high. In normal conditions incentive pricing could be used to encourage peasants to grow a select group of crops if these are considered to be advantageous to the peasantry.

Some may argue against this plan on the grounds that the country has never been able to feed all its people, especially in times of food shortage, and the system will thus not work. We are talking here about relative food surplus and not absolute food surplus. It has always been the case that famine strikes selectively, i.e. some regions or zones are in distress while others are stress-free. Moreover, some areas—like Arssi and Gojjam and central Shoa—are food surplus areas. Now, the national food reserve system will function by stockpiling food during years of good harvest (at a modest rate), and locating this food strategically so that it can be released in distressed areas quickly. It is also often the case that a sharp rise in food prices in crisis areas occurs only for a short time, thus a strategic intervention by a food reserve agency at a critical point will be enough to stabilize prices. The food should be released in a controlled fashion, and to make it difficult for profiteers and speculators to either hoard grain or obtain high windfall profits.

We believe AMC is better equipped than RRC to handle a fullscale strategic food reserve system because of its large network of food stores, purchasing points, and fleet of vehicles. At present, a national food reserve is operated by RRC, but this is based on food aid, and not on local production. The system we are suggesting should be able to rely on local sources for a good part of its needs, and food aid could be made to play a secondary role. The system requires the ability to deploy whatever surplus the country has in strategic areas and as quickly as possible. The food stockpiled for emergency should be collected over a period of many years, mostly from the food surplus areas of the country.

Consider the food that was pumped out of Wollo between 1979, 1982, i.e. the years before the famine (Table 30). If this food (about 40 thousand metric tons), or even half of it was held within the province and released in a controlled manner to the peasantry, thousands of lives would have been saved. The task of feeding the cities should be left to the grain merchant and to private endeavour.

The key element in the reserve system we are suggesting is not so much the volume of food that is collected and in storage, as the effectiveness of quick action: the system hinges on the efficacity of strategic intervention in a crisis, even though this intervention is made with limited resources.

We now turn to what we wish to call *strategic livestock preserve*. This is a system that like the strategic reserve discussed earlier relies on strategic intervention at a critical moment in a crisis. We saw in the previous chapter that perhaps about a quarter of the livestock returned to the famine zones thanks to peasants and rural livestock traders who bought the animals at rock bottom prices, took them out of the famine zone for the duration of the crisis, and brought them back to sell to the peasantry at the end. These smart individuals actually made a strategic intervention in the crisis, and there is no reason why a similar intervention cannot be made on a large-scale, and in an organized and controlled manner. A government, para-government, or independent agency could be empowered to buy cattle and other animals from the peasantry during the crisis, take the animals out of the famine zone until the crisis is over, and then bring them back to sell to the peasantry. The animals should be sold at subsidized or controlled prices as part of a plan to provide rehabilitation assistance to the affected region.

The great advantage of this system is that it will offer the peasantry reasonable prices for their livestock, it will save the animals from death by starvation, and it will make livestock available to the peasantry immediately after the crisis at prices they can afford. This, together with the food reserve system discussed above should serve as an effective weapon against famine. The merit of this package of proposals is that it is an indigenous self-support system, and government and relief agencies need not have to depend on foreign assistance, and emergency famine response need not be paralyzed by the politics of food aid.

We have so far been talking about defensive measures aimed at forestalling danger or minimizing it. Let us deal now with famine response of a different kind, having to do with employment and *employment* creating opportunities, particularly for the rural population.

In the long-run, the structural problems of rural production will be solved not by rural society itself but by the rapid development of industrialization, and the expansion of employment opportunities in the modern sector. This will absorb the excess population from the rural areas, and provide opportunities to the many smallholder peasants a source of livelihood other than unprofitable and inefficient small-scale agriculture. In our opinion government should concentrate most of its efforts and resources on expanding employment in the modern sector, and developing industrialization and support services, however, until the goals of such a programme are achieved, government policy should give maximum encouragement to private small-scale enterprises and cottage industry, both rural and urban.

Paradoxical as it may seem, large-scale industry will in the short-run offer less employment opportunities than small-scale industry and the informal sector. The latter will in this period continue to be the main source of employment. However, the formal sector, once sufficiently developed, should provide more employment in the long-run (see also ILO 1986 C: Chs. 5 and 6).

We have seen above that rural cottage industry in Wollo is fairly underdeveloped in comparison to other parts of the country. Additionally, some sectors of the peasantry here hold unfavourable attitudes about certain crafts and craftsmen—leather-making and smithing are cases in point—and this will have to change. Attitudinal change among the peasantry will come, not as a result of a moral campaign but by making handicraft activity an economically worthwhile endeavour. The expansion of rural cottage industry will be a difficult thing to achieve, and we have no specific proposals on how this can be done, but a determined effort should be made by government, aid agencies, and NGOs to encourage peasants to engage in craft production of one kind or another. This source of supplementary income, especially that which is not seriously affected by environmental stress, could make the difference between life and death in food crisis situations.

Additionally, peasants, through their Associations, could be helped to set up and operate small-scale enterprises that process local products. There is no reason, for example, why with so much oil crops available in many parts of Wollo, oil presses cannot be installed in the rural areas and operated by PAs. Such presses should be labour intensive, using preferably animal power.

Small-scale enterprises, in particular those that use local products and have linkages with agriculture should also be encouraged in the woreda and awraja towns of the province. Enterprises which use livestock and livestock by-products, local crops and local forest products should serve as a source of income to the peasantry.

The fate of the Ethiopian peasant has always been either destitution or death by starvation, and the evidence now being uncovered suggests that the intensity of the first, and the frequency of the second has been growing in the last four decades. If this sorry and tragic record is to be brought to an end, a new, flexible intelligent policy of famine response, one that incorporates the experiences of the peasant, and benefits from the survival strategies evolved by the people themselves, should be designed for each of the major agro-ecological zones of the country and put into effect.

References

1. Information from the Records of:

Ethiopian Red Cross Society, Wollo Province, Dessie, and Bati; Head Office, Addis Ababa.

HASIDA, Wollo Province, Dessie.

Ministry of Agriculture (MOA), Wollo Province, Dessie; Ambassel Awraja, Haiq; Ambassel Woreda, Wichalae; Worebabo Woreda, Bistima; Tehulederae Woreda, Sullula.

Ministry of Health, Wollo Province, Dessie.

Peasant Association, Ambassel Woreda, Wichalae; Worebabo Woreda, Bistima; Tehulederae Woreda, Sullula.

Relief and Rehabilitation Commission, Ambassel Awraja, Haiq.

Save the Children Fund (U.K.), Worebabo Woreda, Bistima.

2. Primary Sources

Addis Zemen. Government owned Amharic language daily Newspaper. Various issues for the year 1984–85.

AMC, *Report on Plan Implementation for the Budget Year 1984/85* (Amharic). Addis Ababa, Hamle 1977 (June 1985).

–, *A Short Study Report on Improving Grain Price Rates* (Amharic). Nehassie 1977 (July 1985).

Australian Agricultural Consulting and Management (AACM), *Ethiopia: Livestock Sub-sector Review. Draft Final. Main Report*, and *Annexes*. Prepared for the Ethiopian Government, Ministry of Agriculture. Adelaide, S. Australia, February 1984.

CSO, *Census Supplement 1*. Addis Ababa, December 1985.

–, *Time Series Data on Area, Production and Yield of Principal Crops by Regions 1979/80–1983/84*. Addis Ababa, December 1984.

Economic Commission for Africa, *Drought and Desertification in Africa*. Special issue, *Rural Progress*, Vol. V, No. 2, 1986.

Ethiopian Meteorological Service, *Meteorological Maps of Ethiopia*. Addis Ababa, 1979.

ENI, *Profile of Wollo under Famine*. ENI 24/74. Addis Ababa, 1984.

–, *Food and Nutrition Situation Prospects for Wollo in 1967* (E.C.) Third Surveillance, 11th October to 3rd November 1974. ENI 114/74. Addis Ababa, 1974.

–, *Guidelines on Feeding Methods for Victims of Natural Disaster* (Amharic). Addis Ababa, n.d. (? 1976).

–, *Report on Baseline Survey Joint WHO/UNICEF Nutrition Support Programme Ethiopia. Part 2*. Addis Ababa, April 1986.

Ethiopian Orthodox Church, *Annual Publication of the Children's Council, No. 2* (Amharic). Addis Ababa, Tiqmit 1978 (October 1986).

Ethiopian Red Cross Society, *Ethiopia: Upper Mille and Cheleka Catchments Disaster Prevention Program. Vol. I: Project Plans and Budget. Vol. II: Land and Water Development Maps. Vol. III: Socio-Economic Features. Vol. IV: Summary and Recommendations*. Addis Ababa: Ethiopian Red Cross Society 1986.

E.S.P., *Ethiopian Serategnoch Party Central Committee Third Regular Meeting* (Amharic). Mimeo, Addis Ababa, 29 Nehassie 1977 E.C. (14 September 1985).

EPID, *Short-Term Recovery Programme for the Drought Stricken Provinces of Ethiopia 1974*. Publication No. 16. Addis Ababa, December 1973.

FAO, *Report of FAO Workshop on National Preparedness for Acute and Large-Scale Food Shortages in Eastern and Southern Africa*. Addis Ababa, 22–26 October 1984.

–, *Natural Resources and the Human Environment for Food and Agriculture in Africa*. Rome: FAO 1986.

FAO/UNDP, *Assistance to Land-Use Planning, Ethiopia. Agroclimatic Resources Inventory for Land-Use Planning*. Vol. I–III, Technical Report 2. Prepared for the Government of Ethiopia. Rome: FAO 1984.

–, *Assistance to Land-Use Planning, Ethiopia. Land-Use, Productive Regions, and Farming System Inventory*. Technical Report 3. Prepared for the Government of Ethiopia. Rome: FAO 1984.

FAO/UNDP, *Assistance to Land-Use Planning, Ethiopia. Economic Analysis of Land-Use*. Technical Report 8. Prepared for the Government of Ethiopia. Rome: FAO 1984.

HASIDA, *Report on a Survey of Small-Scale Industries in Eleven Towns 1976–77*. Addis Ababa, 1979.

–, *Report on a Survey of Private Manufacturing and Non-Manufacturing Engineering Service Industries*. Addis Ababa, April 1985.

ILO, *Socialism from the Grass Roots: Accumulation, Employment and Equity*. Report Prepared by the Employment Advisory Mission. Vol. I: *Report*, Vol. II: *Working Papers*. Addis Ababa: ILO/JASPA, September 1982.

–, *After the Famine. A Programme of Action to Strengthen the Survival Strategies of Affected Populations*. A Report to the Republic of the Sudan. Geneva: ILO 1986, A.

–, *From Crisis to Sustained Development, A Programme of Action.*

ILO, Multi-Disciplinary Programming Mission to Ethiopia. Geneva: ILO 1986, B.

–, *Youth Employment and Youth Employment Programmes in Africa. A Comparative Sub-Regional Study. The Case of Ethiopia*. Addis Ababa: ILO/JASPA 1986, C.

Livestock and Meat Board (LMB), *A Report on the Central Highland Livestock Market Survey*. Prepared by Ayele Gabre Mariam and M. Hillmann. Addis Ababa, August 1975.

Mermarri-Komisyou (Commission of Inquiry): *The Drought and Consequent Famine that Occurred in Wollo Province* (Amharic). Addis Ababa, October 1975.

Ministry of Agriculture (MOA), *Area, Production and Yield of Major Crops for the Whole Country and by Region in 1974/75–1978/79*. Addis Ababa, July 1979.

–, *General Agriculture/Survey, Preliminary Report 1983/84*, Vol. II, Wello. Addis Ababa, October 1984.

Ministry of State Farms Development, *Towards a Strategy for the Development of State Farms in Ethiopia*, Vol. I. Addis Ababa, 1986.

MOA-CSO, *Agricultural Sample Survey 1979/80*, Vol. II, Addis Ababa, May 1980.

MOA-FAO, *Ethiopian Highlands Reclamation Study. A Tentative Review of Agriculture in the Highlands. Working Paper 1.* Prepared by C. Tagoe. Addis Ababa, November 1983.

–, *Ethiopian Highlands Reclamation Study. Report on the Sociological Survey and Sociological Considerations in Preparing a Development Strategy. Working Paper 4.* Prepared by Yeraswork Admassie, et al. Addis Ababa, December 1983.

MOA-FAO, *Ethiopian Highlands Reclamation Study. Agriculture—Strategies for Reclamation and Development. Working Paper 14.* Prepared by P. Cloutier. Addis Ababa, December 1984, A.

–, *Ethiopian Highlands Reclamation Study. The Degradation of Resources and an Evaluation of Actions to Combat It. Working Paper 19.* Prepared by M. Constable. Addis Ababa, December 1984, B.

–, *Ethiopian Highlands Reclamation Study. Annexes. Working Paper 24.* Prepared by M. Constable. Addis Ababa, May 1985.

MOA (Wollo), Livestock and Fisheries Department, *Report for the Year 1984/85* (Amharic). Dessie, Hamle 1977 (August 1985).

NEEPR, *Plan for 1985/86 Part I: General. Part II: Production, Construction, Urban Development. Part III: Economy and Social Services* (Amharic). Dessie, Megabit 1977 (March 1985).

NEERNDRC, *Action Programme (Draft)* (Amharic). Dessie, Hidar 1977 (December 1984).

–, *From October 1984 to September 1985, A Year's Report.* (Amharic). Dessie, 1985, A.

–, *Six Month Action Implementation Report for Ambassel, Dessie Zuria, Wadla Delanta and Qallu Awrajas. Part I* Ambassel Awraja (Amharic). Dessie, Ginbot 1977 E.C. (May 1985), AA.

–, *Six Month Action Implementation Report for Ambassel, Dessie Zuria, Wadla Delanta and Qallu Awrajas. Part III Qallu Awraja* (Amharic). Dessie, Ginbot 1977 E.C. (May 1985), QA.

NRDC-CPSC, *Assessment of the Implementation of the Six Yearly Plans* (Amharic). Addis Ababa, Tahsas 1977 E.C. (December 1984).

RRC, *Food and Nutrition Surveillance Programme, Disaster Area Assessment. Preliminary Report of Wollo Survey.* Addis Ababa, January 1977.

–, *Food Supply System Report.* Periodic report beginning in 1977.

–, *Food Shortages Survey Report on Wollo (Lasta, Wadla Delanta and Werehimenu Awrajas).* Addis Ababa, May 1978.

–, *Review of the Current Drought Situation in Ethiopia.* Addis Ababa, December 1984.

RRC, *The Challenge of Drought: Ethiopia's Decade of Struggle in Relief and Rehabilitation.* Addis Ababa, 1985, A.

–, *Combatting the Effects of Cyclical Drought in Ethiopia.* Addis Ababa, 1985, B.

–, *May Starvation Cease.* Addis Ababa, 1975.

–, *Review of Drought Relief and Rehabilitation Activities for the Period December 1984–August 1985 and 1986 Assistance Requirements.* Addis Ababa, October 1985.

–, *Review of Assistance Requirements (1986/87).* Addis Ababa, January 1987.

RRC, Dawit Wolde Giorgis, Chief Commissioner: Institutional and Practical Responses to Drought in Ethiopia. Address delivered at the 8th International Conference of Ethiopian Studies, Addis Ababa, 29 November 1984.

United Nations Coordinating Committee for Relief and Rehabilitation, *Short-term Relief and Rehabilitations Needs in Ethiopia.* A Report prepared for RRC. Addis Ababa, March 1981.

U.N. Emergency Prevention and Preparedness Group in Ethiopia, Final Draft. Review of Emergency Programmes in Ethiopia in 1986. Unpublished Report, 5 March 1987, Addis Ababa.

U.N. Office for Emergency Operations in Ethiopia, Summary of Bilateral and UN Organizations Contribution to Emergency Operations in Ethiopia (Dec. '84–Dec. '85 Period). Memorandum, 23 October 1985, Addis Ababa.

–, *NGO Newsletter*, Monthly 1985–86. Addis Ababa.

U.S. Congress, House of Representatives, Committee on Foreign Relations, *Human Rights and Food Aid in Ethiopia*. Hearing before the Sub-Committee on Human Rights and International Organization and the Sub-Committee on Africa, 16 October 1985. Washington, D.C.: Government Printing House, 1986.

Wollo Province, Development Office, *Idquetna Limat be-Wollo Teqlai Gizat* (Amharic). Dessie, Yekatit 1961 E.C. (February 1968 G.C.).

World Bank, *Ethiopia, Review of Farmers' Incentives and Agricultural Marketing and Distribution Efficiency*. A Joint Government of Ethiopia and World Bank Mission. N.P.P., March 1983.

–, *Ethiopia, Population, Health, and Nutrition Sector Review*. Report No. 5299—ET. Eastern African Regional Office, 1985.

–, *Ethiopia, Recent Economic Developments and Prospects for Recovery and Growth*. Report No. 5929—ET, Eastern and Southern Africa Region, February 25, 1987.

3. Secondary Sources

Note: Ethiopian authors are listed alphabetically by first name.

Abiye Astatke *et al.*, "Effect of Diet restriction on work performance and weight loss of local Zebu and Friesian X Boran crossbred oxen", *ILCA Bulletin No. 23*, January 1986.

(Zemenfes Qidous) Abraha, *Metsahaffe Aboushak'r (The Book of Aboushak'r)* (Amharic). Adua, 1946 E.C. (1954 G.C.), No Publisher.

W.M. Adams, "Traditional Agriculture and Water Use in the Sokoto Valley, Nigeria", *Geographical Journal*, Vol. 152, Part 1, March 1986.

Afeworq Gebre Eyessus, *Dagmawi Menelik Negus's Negest Ze Etyopya* (Amharic). Rome: 1901 (n.p.).

African Environment, Social Issue on the Sahelian Drought, Vol. 1, No. 2, 1975.

–, *African Agriculture: New Problems, Old Solutions?* Vol. II, No. 4, 1977.

Ahmed Ali, Prediction of Food Shortages and the Response System of the RRC. Paper presented to the Workshop on Drought in Kenya, Nairobi, May 1986.

M. Alamgir, *Famine in South Asia*. Cambridge, Mass.: O.G. & H. Publishers 1980.

Alula Abate *et al.*, *Evaluation of the Impact of UNICEF Assisted Water Supply Projects in Four Administrative Regions: Bale, Harerge, Shewa, Wello Programme Cycle 1980–1983*. Addis Ababa: Institute of Development Research, August 1986.

Frank A. Anderson, Draught Animal Power Systems in Sub-Saharan Africa: Their Production Impact and Research Needs. In Copland (ed.) 1985.

"Anonymous", "Drought and Famine Relief in Ethiopia", *Disasters*, Vol. 7, No. 3, 1983.

Araya Redda, "The Famine in Northern Ethiopia", *Review of African Political Economy*, No. 27/28, 1983.

B. Ashton, *et al.*, "Famine in China, 1958–61", *Population and Development Review*, Vol. 10, No. 4, 1984.

Asnake Ali, "The Conditions of the Peasantry in Wollo 1919–1935: A Review of the Traditions", *Proceedings of the Third Annual Seminar of the Department of History*. Addis Ababa: Addis Ababa University 1986.

S. Aziz (ed.), *Hunger, Politics and Markets, The Real Issues in the Food Crisis*. New York: New York University Press 1975.

Jonathan Baker, *The Rural-Urban Dichotomy in the Developing World. A Case Study from northern Ethiopia*. Oslo: Norwegian University Press 1986.

L. Bartels, Oromo Religion. Berlin: D. Reimer, 1983.

D.R. Basu, "Sen's Analysis of Famine: A critique", *Journal of Development Studies*, Vol. 22, No. 3, 1986.

Berhanou Abebe, "Hamina: Distique Amhariques", *Annales d'Ethiopie* (Addis Ababa), Tome 8, 1970.

Berhanou Zerihun, *Ma'a bel, Ye-Abyot Wazaema*. Addis Ababa: Ethiopian Book Centre 1980. (An Amharic Novel about the Wollo Famine of 1974.)

H. Bernestein, "African Peasantries: A Theoretical Framework", *Journal of Peasant Studies*, Vol. 6, No. 4, 1979.

Sara Berry, "The Food Crisis and Agrarian Change in Africa: A Review Essay", *African Studies Review*, Vol. 27, No. 2, 1984.

B.M. Bhatia, *India's Food Problem and Policy since Independence*. Bombay: Somaiya Publications 1970.

J.A. Binns, "After the Drought: Field Observations from Mali and Burkina Faso", *Geography*, Vol. 71, June 1986.

S. Birks, "The Mountain Pastoralists of the Sultanate of Oman: Reactions to Drought", *Development and Change*, Vol. 9, No. 1, 1978.

Lars Bondestam, "The Politics of Food on the Periphery", in Harle (ed.).

P. Bowbrick, "The Causes of Famine: A Refutation of Professor Sen's Theory", *Food Policy*, Vol. 11, No. 2, 1986.

–, "Rejoinder: An Untenable Hypothesis on the Causes of Famine", *Food Policy*, Vol. 12, No. 1, 1987.

Jan Brøgger, *Relief and Experience among the Sidamo. A Case Study towards an Anthropology of Knowledge*. Oslo: Norwegian University Press, 1986.

John Caldwell, *et al.*, "Periodic High Risk as a Cause of Fertility Decline in a Changing Rural Environment: Survival Strategies in the 1980–1983 South Indian Drought", *Economic Development and Cultural Change*, Vol. 34, No. 4, 1986.

R. Chambers and R. Longhurst, "Trees, Seasons and the Poor", *IDS Bulletin*, Vol. 17, No. 3, 1986.

Lincoln Chen, *et al.*, "Sex Bias in the Family Allocation of Food and Health Care in Rural Bangladesh", *Population and Development Review*, Vol. 7, No. 1, 1981.

John Clarke, *Resettlement and Rehabilitation, Ethiopia's Campaign against Famine*. London: Harney and Jones Ltd., n.d. (? 1986).

Jason Clay and B.K. Holcomb, *Politics and the Ethiopian Famine 1984–1985*. Cambridge, Ma.: Cultural Survival, December 1985.

Lionel Cliffe, "Capitalism or Feudalism? The Famine of Ethiopia", *Review of African Political Economy*, No. 1, 1974.

Lionel Cliffe and R. Moorsom, "Rural Class Formation and Ecological Collapse in Botswana", *Review of African Political Economy*, No. 15/16, 1979.

Stephen Commins, *et al.* (eds.): *Africa's Agrarian Crisis: The Roots of Famine.* Boulder, Colo.: Lynne Reinner 1986.

J.W. Copland (ed.), *Draught Animal Power for Production.* Proceedings of an International Workshop held at James Cook University, Townsville, Qld, Australia, 10–16 July 1985. Canberra: Australian Centre for International Agricultural Research 1985.

Noel Cossins, *No Way to Live. A Study of the Afar Clans of Northeast Range-lands of Ethiopia.* Addis Ababa: Livestock and Meat Board, 1972.

–, *Green Heat of a Dying Land. A Study of the New Cotton Wealth of the Old Afar Sultanate of Aussa.* Unpublished Ms., Addis Ababa, 1973.

–, (with Bekele Yemerou), *Still Sleep the Highlands. A Study of Farm and Livestock Systems in the Central Highlands of Ethiopia.* Addis Ababa: Livestock and Meat Board 1974.

–, Wollo, the Drought is over but Hunger Lingers on. Report prepared for RRC, Addis Ababa, n.d. (? 1975).

J.T. Crowe and Shitaye Gebre Medhin, *Crop Pesthandbook.* Addis Ababa: Institute of Agricultural Research, June 1977, (Revised edn.).

Peter Cutler, "The Use of Early Indicators of Food Emergencies. Case Study: Ethiopia". Unpublished report, *Food Emergencies Research Unit,* London School of Hygiene and Tropical Medicine, June 1983.

–, "Famine Forecasting: Prices and Peasant Behaviour in northern Ethiopia", *Disasters,* Vol. 8, No. 1, 1984.

Peter Cutler and R. Stephenson, *The State of Food Emergency Preparedness in Ethiopia.* London: International Disaster Institute 1984.

G. Dahl and A. Hjort, *Pastoral Change and the Role of Drought.* Report R2. Stockholm: SAREC 1979.

Daniel Gemechu, "Climate and Crop Production in Wollo", in *Food Crisis in Ethiopia,* a forthcoming monograph to be published by CODESRIA, Dakar.

David Dalby *et al.* (eds.), *Drought in Africa 2.* African Environment Special Report 6. London: International African Institute 1977.

Department of Biology, Science Faculty, A Study on the Medicinal Plants of Ethiopia. Part 1A, the Identity of some of the Plants, Addis Ababa University, MS., n.d.

J. Derrick, "The Great West African Drought, 1972–74", *African Affairs,* Vol. 76, No. 305, 1977.

–, "West Africa's Worst Year of Famine", *African Affairs,* Vol. 83, No. 332, 1984.

Dessalegn Rahmato, *Agrarian Reform in Ethiopia.* Trenton, N.J.: Red Sea Press 1984.

–, "Moral Crusaders and Incipient Capitalists: Mechanized Agriculture and its Critics in Ethiopia", *Proceedings of the Third Annual Seminar of the Department of History.* Addis Ababa: Addis Ababa University Press 1986 A.

–, "The Crisis of Livelihood in Ethiopia". Paper prepared for the Peace and Developed Panel of the *International Peace Research Conference,* Brighton, University of Sussex, 13–18 April 1986 B.

–, "Food Aid and Food Dependency in Ethiopia", Paper presented at the *OSSREA Congress,* Eldoret, Kenya, 28–31 July 1986 C.

–, "Some Notes on Settlement and Resettlement in Mettekel Awraja (Gojjam Province)". Paper presented at the *9th International Conference of Ethiopian Studies, Moscow, 26–31 August 1986 D.*

D. Devitt, "Coping with Drought in the Kalahari", in Dalby *et al.* (eds.).

Barbara Dinham and C. Hines, *Agribusiness in Africa*. Trenton, N.J.: Africa World Press 1984.

R. Dirks, "Social Response during Severe Food Shortages and Famine", *Current Anthropology*, Vol. 21, No. 1, 1980.

R. Dumont and No. Cohen, *The Growth of Hunger, A New Politics of Agriculture*. London: Marion Boyars 1980.

C.M. Edmondson, "The Politics of Hunger: The Soviet Response to Famine, 1921", *Soviet Studies*, Vol. 29, No. 4, 1977.

Fassil G. Kiros and Alemayehu Lirenso, "Analysis of Grain Marketing and Pricing Policies in Ethiopia with Particular Reference to the Staple Teff". Unpublished Ms., *Institute of Development Research*, Addis Ababa, November 1985.

R.H. Faulkingham, "Ecological Constraints and Subsistence Strategies: The Impact of Drought in a Hausa Village, A Case Study from Niger", in Dalby, *et al.* (eds.).

Fikarae Yesus, Date of Original Ms. unknown. Amharic translation published by Tesfa Gebre-Selassie in 1964 E.C. (1972 G.C.).

J. Firebrace and S. Holland, *Never Kneel Down. Drought, Development and Liberation in Eritrea*. Trenton, N.J.: Red Sea Press 1985.

Harold Fisher, *The Famine in Soviet Russia 1919–1923*. Freeport, N.Y.: Books for Libraries 1927.

M. Fitzgerald, *Drought, Famine and Revolution in Ethiopia*. Occasional Paper 1, Department of Geography. London: University of London 1980.

Tom Franklin, Cleft Stick, *Dilemmas in the Development of a Disaster: Ethiopia 1974–1984*. (Ms. completed 1987).

–, "The Build Up to the 1984 Ethiopian Famine". (A Chronology) Addis Ababa, Ms., 1986.

Harriet Friedmann, "The Political Economy of Food, The Rise and Fall of the Postwar International Food Order", in M. Burawoy and T. Skocpol (eds.), *Marxist Enquiries*, Chicago: University of Chicago Press 1982.

G. Galperin, "Ethiopia, Stages in Agrarian Revolution", Africa and Asia Today (Moscow), No. 6, 1981.

–, "Ethiopia: Aid, Facts and Fiction", *New Times* (Moscow), No. 2, 1985.

Beverly Gartrell, "Searching for the 'Roots of Famine': The Case of Karamoja", *Review of Africa Political Economy*, No. 33, 1985.

Susan George, *How the Other Half Dies*. Baltimore, Md.: Penguin Books 1976.

–, *Ill Fares the Land*. London: Writers and Readers 1984.

Peter Gill, *A Year in the Death of Africa. Politics, Bureaucracy and the Famine*. London: Paladin Books 1986.

B.A. Godana, "The Inter-Governmental Authority on Drought and Development in Eastern Africa". Paper presented at the OSSREA Congress, Eldoret, Kenya, 28–31 July, 1986.

Reginald Green, "Food Policy, Food Production, and Hunger in Sub-Saharan Africa: Retrospect and Prospect", *International Journal*, Autumn 1986.

P.R. Greenough, "Indian Famines and Peasant Victims: The Case of Bengal in 1943–44", *Modern Asian Studies*, Vol. 14, Pt. 2, 1980.

G. Gryseels, "Livestock in Farming Systems Research for Smallholder Agriculture. Experiences of ILCA's Highland Programme". Paper presented at the Seminar on *Agricultural Research in Rwanda*, Kigali, 5–12 February 1983.

G. Gryseels and F.M. Anderson, *Research on Farm and Livestock Productivity in the Central Ethiopian Highlands: Initial Results 1977–1980.* Research Report No. 4 Addis Ababa: ILCA December 1983.

G. Gryseels and M.R. Goe, "Energy Flows on Smallholder Farms in the Ethiopian Highlands", *ILCA Bulletin* No. 17, January 1984.

G. Gryseels and S. Jutzi, *Regenerating Farming Systems after Drought: ILCA's Ox/Seed Project, 1985 Results.* Addis Ababa: ILCA July 1986.

G. Gryseels *et al.:* "The Use of Single Oxen for Crop Cultivation in Ethiopia", *ILCA Bulletin,* No. 18, April 1984.

Hailu Lemma, "The Politics of Famine in Ethiopia", *Review of African Political Economy,* No. 33, 1985.

C.R. Hallpike, *The Konso of Ethiopia.* Oxford: Clarendon Press 1972.

G. Hancock, *Ethiopia: The Challenge of Hunger.* London: Gollancz 1985.

Dag Hareide, "Food for Work in Ethiopia. A Short Introduction to who is doing what, where, when, how and with whom. Paper presented at the Workshop hosted by the National Committee for Central Planning, Addis Ababa, 21 July 1986.

V. Harle (ed.), *The Political Economy of Food.* Farnborough: Saxon House 1978.

P. Harrison and R. Palmer, *News out of Africa.* London: Hilary Shipman 1986.

B.L. Henricksen and J.W. Durkin, "Moisture Availability, Cropping Period and the Prospects for Early Warning of Famine in Ethiopia", *ILCA Bulletin* No. 21, January 1985.

Paul Henze, "Behind the Ethiopian Famine, Anatomy of a Revolution", *Encounter,* June, July/August and September–October 1986.

M.T. Hinchey (ed.), *Proceedings of the Symposium on Drought in Botswana 5–8 June 1978.* Gaborone: Botswana Society 1979.

R.R. Hitchcock, "The Traditional Response to Drought in Botswana", in Hinchey (ed.).

Julius Holt and P. Cutler, *Review of the Early Warning System of the Relief and Rehabilitation Commission.* Addis Ababa: RRC/UNICEF, March 1984.

H.P. Huffnagel, *Agriculture in Ethiopia.* Rome: FAO 1961.

G.W.B. Huntingford, *The Galla of Ethiopia.* London: International African Institute 1955.

A. Hussain and K. Tribe, *Marxism and the Agrarian Question.* 2nd. ed. London: MacMillan Press 1983.

A.M. Hussein (ed.), *Rehab. Drought and Famine in Ethiopia.* London: International African Institute 1976.

IDS Bulletin, *Feeding the Hungry: A Role for Post-Harvest Technology?* Special issue, Vol. 13, No. 3, 1982.

–, *Seasonality and Poverty.* Special issue, Vol. 17, No. 3, 1986.

Kurt Jansson *et al., The Ethiopian Famine.* London: Zed Books 1987.

C. Jest, "The Future of Mountain Peasantry", *Ceres,* July–August 1986.

Janice Jiggins, "Women and Seasonality: Coping with Crisis and Calamity", *IDS Bulletin,* Vol. 17, No. 3, 1986.

P. King, *An African Winter*. London: Penguin Books 1986.

H. Kloos, "Development, Drought and Famine in the Awash Valley of Ethiopia", *African Studies Review*, No. 4, 1982.

Karl E. Knutsson, *Authority and Change: A Study of the Kallu Institution among the Macha Galla of Ethiopia*. Göteborg: Etnografiska Museet 1967.

P. Koen, "Ethiopia: Famine, Food Production, and Changes in the Legal Order", *African Studies Review*, Vol. 22, No. 1, 1979.

David Korn, "Ethiopia: Dilemma for the West", *The World Today*, January 1986.

–, *Ethiopia, The United States and the Soviet Union*. London: Croom Helm 1986.

Jocelyn Kynch and A. Sen, "Indian Women: Well-being and Survival", *Cambridge Journal of Economics*, Vol. 7, No. 3–4, 1983.

Frances Lappe and J. Collins, *Food First: The Myth of Scarcity*. London: Souvenir Press 1977.

A. Leftwich and D. Harvie, *The Political Economy of Famine*. Discussion Paper 116, Institute for Research in the Social Sciences. York: University of York 1986.

Norman Long (ed.), *Family and Work in Rural Societies. Perspectives on Non-Wage Labour*. London: Tavistock Publications 1984.

R. Longhurst, "Household Food Strategies in Response to Seasonality and Famine", *IDS Bulletin*, Vol. 17, No. 3, 1986.

K.J. Lundström, *North-Eastern Ethiopia: Society in Famine*. Research Report No. 34. Uppsala: Scandinavian Institute of African Studies 1976.

G.H. Maddox, "NJAA: Food Shortages and Famines in Tanzania between the Wars", *International Journal of African Historical Studies*, Vol. 19, No. 1, 1986.

M. Mamdani, "The Colonial Roots of Famine in Karamoja: A Rejoinder", *Review of African Political Economy*, No. 36, September 1986.

T.J. Manetsch, "On Food Shortage Forecasting with Special Emphasis on the South Asian Case", *The Bangladesh Development Studies*, Vol. V, No. 3, 1977.

M.B. McAlpin, "Dearth, Famine and Risk: The Changing Impact of Crop Failures in Western India, 1870–1920", *Journal of Economic History*, Vol. 39, No. 1, 1979.

Jemes McCann, "Households, Peasants, and Rural History in Lasta, Northern Ethiopia, 1900–35. Unpublished dissertation, University of Michigan, 1984.

–, "Household Economy: Demography, and the 'Push' Factor in Northern Ethiopian History, 1916–35", Review, *Winter* 1986.

–, "The Social Impact of Drought in Ethiopia: Oxen, Households, and Some Implications for Rehabilitation". Unpublished Ms., African Studies Centre, Boston University, 1986.

John Mellor, "Prediction and Prevention of Famine". International Food Policy Research Institute Reprint No. 95. Washington, D.C., 1986.

–, and Sara Gavian: "Famine: Causes, Prevention, and Relief", *Science*, 30 January 1987.

Mengistu Lemma, *Metsehaffe Tizzita ze-Alega Lemma* (Amharic). Addis Ababa: Berhanena Selaam Printing Press 1959 E.C. (1966/67 G.C.).

Merid W. Aregay, "Land Tenure and Agricultural Productivity, 1500–1850", *Proceedings of the Third Annual Seminar of the Department of History*. Addis Ababa University 1986.

Mesfin Wolde Mariam, *Rural Vulnerability to Famine in Ethiopia 1958–1977*. New Delhi: Vikas Publishers 1984.
–, "Northern Shewa and Wello, Background Paper on a Development Strategy for the Problem of Vulnerability to Famine". Report prepared for FAO Investment Centre, FAO, Addis Ababa, September 1985.
–, "Ethiopia's Food Security: Problems and Prospects". Prepared for the National Workshop on Food Strategies for Ethiopia, Alemaya University of Agriculture, Alemaya, 8–13 December 1986.
Mogues Azbite, "A Famine Relief Operation at Qorem, Ethiopia in 1966", *Disasters*, Vol. 5, No. 1, 1981.
H.F. Mooney, *A Glossary of Ethiopian Plant Names*. Dublin: Dublin University Press 1963.
M. Muqtada, "Poverty and Famine in Bangladesh", Vol. IX, No. 1, 1981. *Bangladesh Development Studies*.
National Research Council, Committee on *International Disasters Needs*. Washington, D.C.: National Academy of Sciences 1979.
James Newman (ed.), *Drought, Famine and Population Movements in Africa*. Syracuse N.Y.: Maxwell School of Citizenship and Public Affairs, Syracuse University 1975.
Peter Niggli, *Ethiopia: Deportation and Forced Labour Camps*. Berlin: Berliner Missionwerk 1986.
D.W. Norman, "The Rationalisation of Intercropping", *African Environment*, Vol. II, no. 4, Nov. 1977.
Alec Nove, *An Economic History of the USSR*. Harmondsworth: Penguin Books, 1976.
Jay O'Brien, "Sowing the Seeds of Famine: The Political Economy of Food Deficits in the Sudan", *Review of African Political Economy*, No. 33, 1985.
P. O'Keefe and B. Wiener (eds.), *Land-Use and Development*. African Environment Special Report. 5. London: International African Institute 1977.
Gail Omvedt, "The Political Economy of Starvation", *Race and Class*, Vol. 17, No. 2, 1975.
Oxfam, *Lessons to be Learned. Drought and Famine in Ethiopia*. Oxford: Oxfam 1984.
–, *Sudan, The Roots of Famine*. A Report by Nick Carter. Oxford: Oxfam 1986.
Richard Pankhurst, *The History of Famine and Epidemics in Ethiopia Prior to the Twentieth Century*. Addis Ababa: RRC 1985.
–, *Economic History of Ethiopia 1800–1935*. Addis Ababa: Haile Selassie University Press 1968.
M. Peberdy, *Tigray: Ethiopia's Untold Story*. London: Relief Society of Tigray 1985.
M. Pelly, Article (untitled). *The Lancet*. London 17 May 1986.
J.P. Pottier, "The Politics of Famine Prevention: Ecology, Regional Production and Food Complementarity in Western Rwanda", *African Affairs*, Vol. 85, No. 339, 1986.
M. Ravaillion, "The Performance of the Rice Market in Bangladesh during the 1974 Famine", *Economic Journal*, 95, March 1985.
M. Redclift, "Sustainability and the Market: Survival Strategies in the Bolivian Frontier", *Journal of Development Studies*, Vol. 23, No. 1, 1986.
Review of African Political Economy, *The Roots of Famine*. Special issue, No. 15/16, 1979.
–, *War and Famine*. Special Issue, No. 33, 1985.
Review of African Political Economy, "Nations and Nationalities in Ethiopia and Eritrea", (Documents from EPLF and TPLF). No. 35, May 1986.

Paul Richards, "'Alternative' Strategies for African Environment". 'Folk Ecology' as a Basis for Community Oriented Agricultural Development", in Paul Richards (ed.).

–, (ed.), African Environment: *Problems and Prospects.* African Environment Special Report 1. London: International African Institute 1975.

M. Rosenzweig and T. Shultz, "Market Opportunities, Genetic Endowments and Intra-family Resource Allocation: Child Survival in Rural India", *American Economic Journal*, Vol. 72, No. 4, 1983.

R.I. Rotberg (ed.), *Imperialism, Colonialism and Hunger: East and Central Africa.* Lexington, Ma.: D.C. Health and Co. 1983.

Robert Roundy, "Altitudinal Mobility and Disease Hazards for Ethiopian Populations, *Economic Geography*, Vol. 25, No. 2, 1976.

A. Salifou, "When History Repeats Itself: The Famine of 1931 in Niger", *African Environment*, Vol. 1, No. 2, 1975.

S. Sandford, *Organization and Management of Water Supplies in Tropical Africa.* Addis Ababa: ILCA, December 1983.

Save the Children Fund (U.K.), *Wollo Nutrition Field Worker Program.* Monthly Reports (for the Year 1986) Dessie, 1986.

B.G. Schoepf, "Food Crisis and Class Formation in Shaba", *Review of African Political Economy.* No. 33, 1985.

D.J. Schove, "African Droughts and the Spectrum of Time", in Dalby *et al.* (eds.).

James Scott, *The Moral Economy of the Peasant.* New Haven: Yale University Press 1976.

John Seaman and J. Holt, *Harerghe under Drought. A Survey of the Effects of Drought upon Human Nutrition in Harerghe Province.* Report prepared for RRC. Addis Ababa: RRC, May/June 1974.

Janet Seeley, *Famine in Africa: A Guide to Bibliographies and Resource Centres.* Cambridge: African Studies Centre, University of Cambridge 1986 A.

–, *Famine in Sub-Saharan Africa: A Selected Bibliography (excluding the Sahel) from 1978.* Cambridge: African Studies Centre, University of Cambridge 1986 B.

Ruth Selinus, *The Traditional Foods of the Central Ethiopian Highlands.* Research Report No. 7, Uppsala: Scandinavian Institute of African Studies, 1971.

A.K. Sen, "Famines", *World Development*, Vol. 8, No. 9, 1980.

–, "Ingredients of Famine Analysis. Availability and Entitlement", *Quarterly Journal of Economics*, Vol. XCVI, No. 3, 1981 A.

–, *Poverty and Famines: An Essay on Entitlement and Deprivation.* Oxford: Clarendon Press 1981 B.

–, "The Battle to Get Food", *New Society*, 13 October 1983.

–, "The Causes of Famine. A Reply", *Food Policy*, 01. 11, No. 2, 1986, Vol. 12, No. 1, 1987.

William Shack, "Hunger, Anxiety and Ritual: Deprivation and Spirit Possession among the Gurage of Ethiopia", *Man*, Vol. 6, No. 1, 1971.

B. Shenton and M. Watts, "Capitalism and Hunger in Northern Nigeria", *Review of African Political Economy*, No. 15/16, 1979.

Jack Shepherd, *The Politics of Starvation.* New York: Carnegie Endowment for International Peace 1975.

–, "Politics and Bureaucracy Let This Famine Happen", *Washington Post*, 18 November 1984.

–, "Ethiopia: The Use of Food as an Instrument of U.S. Foreign Policy", *Issue, a Journal of Opinion*, Vol. IV, 1985.

–, "Ethiopia's Famine: The Politics of Food Aid", *Africa Report*, March–April 1985.

Thomas Shopo, "The Political Economy of Hunger", I. Mandaza, ed., *Zimbabwe, The Political Economy of Transition 1980–86*, Dakar, CODESRIA Books Series 1987.

G. Sivini, "Famine and the Resettlement Program in Ethiopia", *Africa* (Rome, Anno XLI) No. 2 1986.

P. Spitz, "Silent Violence: Famine and Inequality", *International Social Science Journal*, Vol. 30, No. 4, 1978.

T.N. Srinivasan, Hunger: Defining it, Estimating its Global Incidence, and Alleviating it. In G. Johnson and G. Schuh (eds.). The Role of Markets in the World Food Economy. Boulder: Westview Press 1983.

S. Strelcyn, *Medecine et Plantes d'Ethiopie II Enquête sur les Noms et l'Emploi des Plantes en Ethiopie*. Napoli: Instituto Universitario Orientale 1973.

Survival International, *Ethiopia's Bitter Medicine. Settling for Disaster: An Evaluation of the Ethiopian Government's Programme*. London: Survival International 1986.

R.H. Tawney, *Land and Labour in China*. Boston: Beacon Press 1966 (1932 edn.).

A.J. Taylor, "Mixed Cropping as an Impact on the Management of Crops Pests in Tropical Africa", *African Environment*, Vol. II, No. 4, No. 5, 1977.

Teclehaimanot G. Selassie, "Way to Response to Food Shortages", *Proceedings of the Third Annual Seminar of the Department of History*. Addis Ababa University 1986.

L. Timberlake, *Africa in Crisis. The Causes, the Cures of Environmental Degradation*. London: Earthscan 1985.

A.R. Tippet, Peoples of Southwest Ethiopia. Pasadena: William Carey 1970.

J.H. Topps, "Adaptation of Cattle to Drought conditions and their Requirements for Food and Water", in Dalby *et al.* (eds.).

William Torry, "Drought and the Government—Village Emergency Food Distribution System in India", *Human Organization*, Vol. 45, No. 1, 1986.

Camilla Toulmin, *Livestock Losses and Post-Drought Rehabilitation in Sub-Saharan Africa*. LPU Working Paper No. 9. Addis Ababa: ILCA, December 1985.

–, "Pastoral Livestock Losses and Post-Drought Rehabilitation in Sub-Saharan Africa: Policy Options and Issues", *Alpan* (ILCA), Network Paper No. 8, March 1986.

–, "Drought and the Farming Sector: Loss of Farm Animals and Post-Drought Rehabilitation", *Alpan* (ILCA), Network Paper No. 10, September 1986.

J.S. Trimingham, *Islam in Ethiopia*. London: Frank Cass 1965.

David Turton, "Mursi Response to Drought: Some Lessons for Relief and Rehabilitation", *African Affairs*, Vol. 84, July 1985.

–, and P. Turton, "Spontaneous Resettlement after Drought: An Ethiopian Example, *Disasters*, Vol. 8, No. 3, 1984.

Nigel Twose, *Cultivating Hunger. An Oxfam Study of Food, Power and Poverty*. Oxford: Oxfam 1984.

–, and M. Goldwater: *Fighting the Famine*, London: Pluto Press 1985.

Megan Vaughan, "Famine Analysis and Family Relations: 1949 in Nyasaland", *Past and Present*, No. 108, August 1985.

T.M. Vestal, "Ethiopia's Famine: A Many-Dimensioned Crisis", *The World Today*, July 1985.

Michael Watts, *Silent Violence: Food, Famine and Peasantry in Northern Nigeria.* Berkeley: University of California Press, 1983.

E. Westphal, *Agricultural Systems in Ethiopia.* Wageningen: Centre for Agricultural Publishing and Documentation 1975.

Ben Wisner, "Man-made Famine in Eastern Kenya: The Inter-relationship on Environment and Development", in O'Keefe and Wisner (eds.).

Wolde Michael Kelecha, *A Glossary of Ethiopian Plant Names.* 2nd Revised Edition. Addis Ababa, n.p. March 1977.

A.P. Wood, "Farmers Response to Drought in Ethiopia", in Hussein, ed., 1976.

–, "Resettlement in Illubabor Province, Ethiopia. Unpublished dissertation, University of Liverpool, 1977.

C.A. Wood, "A Preliminary Chronology of Ethiopian Droughts", in Dalby *et al.* (eds.).

K. Wright, "Combating Famine: A Revolutionary Strategy", Review of African Political Economy, No. 30, 1984.

Yeraswork Admassie and Solomon Gebre, *Food-for-Work in Ethiopia: A Socio-Economic Survey.* Report prepared for the World Food Programme and the Ministry of Agriculture. Research Report No. 24. Addis Ababa: Institute of Development Research, Addis Ababa University 1985.

Yohannes Gebre Tsadik, "Ethiopia: Where even the Camels are Dying", UNICEF News, Issue 120, 2, 1984.

Annex 1

Rainfall Data, Haiq Station, Ambassel Awraja, Wollo, 1963–1984 (in millimetres)

		Belg (Spring)			Meher (Autumn)		
Year	Annual	March	April	May	July	August	September
1963	1294.2	16.0	223.9	194.1	293.5	184.5	89.0
1964	1673.7	61.3	111.5	68.8	472.7	334.2	70.1
1965	1089.6	47.6	145.0	7.4	236.7	327.9	67.8
1966	1977.5	66.8	189.8	45.9	105.7	250.7	65.5
1967	1236.3	145.4	75.6	188.7	296.5	249.9	51.7
1968	1344.6	98.5	191.2	13.4	500.7	156.4	100.9
1969	1296.6	208.4	93.4	81.1	277.8	299.7	58.6
1970	1259.8	174.4	38.7	17.7	380.5	297.2	162.2
1971	1201.6	93.3	80.1	195.9	261.3	349.3	98.6
1972	913.3	81.3	192.9	48.9	146.0	228.5	35.3
1973	930.1	4.3	31.8	120.4	257.5	270.8	148.7
1974	1199.8	110.3	31.2	125.5	281.9	297.6	207.0
1975	1484.3	63.3	102.6	25.4	233.5	398.8	247.9
1976	1142.2	79.3	230.4	131.7	156.2	212.6	129.4
1977	1122.4	77.4	87.7	82.9	264.3	288.5	72.2
1978	1043.9	119.2	29.2	69.3	254.0	257.7	125.0
1979	1189.9	120.6	72.2	160.0	226.8	227.4	85.5
1980	1160.3	33.9	122.9	35.6	323.0	294.8	87.7
1981	1252.4	277.0	141.9	19.0	395.8	223.6	136.6
1982	1157.9	115.1	207.4	66.9	120.4	263.0	130.1
1983	1235.5	197.6	239.2	161.6	163.1	227.5	76.6
1984	572.8	14.4	22.2	187.3	120.5	33.6	124.3
Mean	1175.0	104.3	116.0	88.3	260.7	261.4	108.7

Source: Daniel Gemechu, Geography Department, Addis Ababa University.

Annex 2

Livestock supply and prices, Haiq Market, Ambassel. Tir 1976 EC (Jan. 1984)–Nehassie 1977 (August 1985)

Livestock	1976 (1984)								
	Tir Jan	Yekatit Feb.	Megabit March		Miazia April	Gunbot May	Sene June	Hamle July	Nehassie August
Vol.[b]	590	816	692		838	586	1391	2138	2007
Ox/Bull									
Price[c]	110	120	105		125	125	50	60	65
Vol.	223	284	372		317	131	370	332	340
Cow/Heifer									
Price	80	85	80		85	105	40	40	40
Vol.	356	344	260		367	120	446	851	780
Calves									
Price	40	55	40		45	35	30	25	25
	1977 EC			1985					
	Mes- kerem Sept.	Tiqimt Oct.	Hidar Nov.	NA[a] –	Miazia April	Gunbot May	Sene June	Hamle July	Nehassie August
Vol.[b]	1849	2426	2896	–	956	1320	1205	763	940
Ox/Bull									
Price[c]	65	55	50	–	125	125	150	150	125
Vol.	510	704	549	–	545	558	663	318	420
Cow/Heifer									
Price	35	40	40	–	75	105	95	75	50
Vol.	1295	1326	937	–	710	614	767	498	569
Calves									
Price	20	20	15	–	35	45	40	50	30

Note: [a] The records are blank for the month of Tahsas to Megabit (Dec. 1984 to March 1985).
[b] The total number of animals at all the market-days in the month. It means the same animal may have appeared at the weekly market more than once.
[c] We have taken the lower price in the month.

Source: Livestock and Livestock Products Marketing Unit, Ministry of Agriculture, Haiq (Ambassel awraja), Registry of Weekly Livestock Market.

Annex 3

Grain prices, Bistima Market (Worebabo woreda), Ambassel. March 1985–February 1986

Grain	No. of Tassa Per Quintal for 100 kg.	Monthly Price in Birr per Tassa											
		1977 EC						1978 EC					
		March Megabit 1985 GC	April Miazia	May Gunbot	June Sene	July Hamel	August Nehassie	September Meskerem	October Tiqimt	Nov. Hidar	Dec. Tahsas	January Tir 1986 GC	February Yekatit
Sorghum	85	2.00	2.00	2.40	2.25	2.40	2.35	2.10	2.25	1.15	1.10	0.75	1.10
Maize	82	1.80	1.95	2.15	1.95	1.90	1.90	1.85	1.25	0.85	0.85	0.95	0.90
Teff	77	2.80	3.20	3.20	2.85	2.60	2.50	2.75	2.50	1.85	1.95	1.90	1.90
Barley	106	1.60	2.00	1.65	1.35	2.40	1.45	1.10	1.25	0.95	0.75	0.75	1.20
Wheat	80	1.95	2.15	2.20	2.00	1.95	1.95	1.95	1.30	1.20	0.95	0.95	1.60
Beans	77	2.00	2.20	2.15	2.25	2.40	2.00	2.10	1.80	1.95	1.25	1.75	1.20
Lentils	72	2.60	2.40	3.20	3.20	3.05	2.85	2.85	2.85	2.20	1.55	1.85	1.75
F/Peas	79	2.05	2.40	NA	2.45	2.80	3.60	3.20	2.10	2.10	1.60	1.95	1.60

Source: Save the Children Fund (U.K.), Bistima Office (November 1986)

Note: A *tassa* is a can for measuring grain in the rural market. Eg. 85 tassas make one quintal of sorghum which at 2.00 Birr per tassa comes to 170 Birr per quintal.

Annex 4

Recipients of relief aid in Wollo (1984 by awraja)

Awraja	Rural Census	Beneficiaries by age group			
		0–6	7–15	Over 15	Total
1. Raya Qabo	145 995	46 494	47 587	112 613	206 694
2. Wag	209 083	26 167	51 925	130 029	208 121
– Yejju	251 247	46 382	50 193	114 706	211 281
3. Lasta	337 942	28 615	4013	71 135	103 766
– Wore-Himenu	343 796	14 703	21 943	57 416	94 062
– Ambassel	350 792	28 178	17 161	50 805	96 144
– Qallu	360 785	15 832	37 718	65 510	122 060
– W/Delanta	243 439	10 288	12 614	37 988	60 890
– Dessie Zuria	303 371	1269	4956	7333	13 558
– Dessie Shelter[a]	–	120	160	116	396
Korem[a]	–	7278	–	5027	12 305
4. Aussa	126 597	8851	8581	16 069	33 501
Total	3.4 m	234 177	292 491	671 747	1 198 415

Note: [a] These two shelters have been listed independently. Korem is in Wag awraja. Wore-Illu and Borena awrajas are not included because they did not suffer famine in 1984. NEERNDRC (1985) does not show any relief recipients from these awrajas in 1985. The population figures for awrajas 1 to 4 are estimates. There must have been some double counting in the relief centre in the first three awrajas. The total figure of 3.4 million is for all of rural Wollo.

Source: NEERNDRC 1984:11. CSO (1985) for Census.

RRC records show that 2.6 million Wollo peasants were affected by famine in December 1984. The agency now says (January 1987) that of the 2.5 million peasants in the country who still need food aid 334,000 are Wolloyaes.

Annex 5

National and International Relief Organizations in Wollo, 1984–1985

Organization	Operation Centre	Awraja
Africare (U.S.)	Woldiya	Yejju
Catholic Relief (U.S.)	Qobbo	Raya Qobbo
Concern (Irish)	Harbu, Kemissie, Chiretti, Bati	Qallu
Ethiopian Orthodox Church	Haiq	Ambassel
	Genet Ber (Gbalaffo)	Yejju
Ethiopian Red Cross	Korem	Wag
	Alamata, Qobbo	Raya
	Lalibella	Lasta
	Sanga	Yejju
	Bistima, Gerba	Ambassel
	Bati, Degan	Qallu
	Millae, Assaita	Aussa
German Emergency Doctors	Lalibella	Lasta
Internation Committee of Red Cross	Many parts of Wollo	
Italian Medical Team	Bati	Qallu
Japan Relief Services (JUC)	Ajjibar	Wor-Himenu
Lutheran World Fed. (U.S.)	Jari	Ambassel
Medicins sans Frontier (France)	Korem	Wag
Mekane Yesus (Ethiopian)	Qobbo	Raya Qobbo
	Mersa	Yejju
	Tis'abalima	Ambassel
Mennonite Mission (U.S.)	Gerado Valley	Dessie Zuria
Mission of Charity (?)	Alamata	Raya Qobbo
Oxfam (U.K.)	Qobbo	Raya Qobbo
	Wogel Tena, Tsehai Mewcha	Wadla Del.
	Bora, Harwa	Qallu
Philadelphia Church Mission (?)	Qundie (Kta Ber)	Dessie Zuria
Rädda Barnen (Swedish)	Many parts of Wollo	
SCF (U.K.)	Korem	Wag
	Qobbo	Raya Qobbo
	Bistima, Bulbullo	Ambassel
World Vision (U.S.)	Alamata	Raya Qobbo
	Lalibella	Lasta
	Sanga	Yejju
	Ajjibar	Wore-Himenu

Note: We have excluded international organizations which worked through local agencies. Eg. the German, Japanese and Swedish Red Cross Societies are not listed because they worked through the Ethiopian Red Cross.

Source: NEERNDRC documents.

Annex 6

Emergency food aid to Ethiopia 1984/85 to 1986 (in '000 US Dollars)

Bilateral Donors	1984/85	1986	Total
Australia	14,474	6,082	20,556
Austria	1,080	3,194	4,274
Belgium	3,970	–	3,970
Bulgaria	5,219	–	5,219
Canada	41,974	20,200	62,174
China	4,082	2,000	6,082
Czechoslovakia	2,375	–	2,375
Denmark	514	166	680
EEC	59,951	72,800	132,751
Finland	2,378	–	2,378
France	23,045	3,200	25,245
East Germany	1,353	–	1,353
West Germany	17,124	9,059	26,183
Greece	4,801	–	4,801
Hungary	1,624	–	1,624
Iceland	51	–	51
Iran	1,392	–	1,392
Ireland	1,232	–	1,232
India	15,390	1,600	16,990
Italy	6,940	5,618	12,558
Japan	6,169	–	6,169
Korea, Rep.	150	–	150
Libya	270	–	270
Holland	5,735	2,110	7,845
New Zealand	38	–	38
Norway	5,611	–	5,611
Pakistan	135	–	135
Poland	3,483	–	3,483
Romania	6,000	–	6,000
Spain	3,500	–	3,500
Sweden	3,891	7,960	11,851
Switzerland	3,020	400	3,420
U.K.	5,469	2,800	8,269
USA	252,069	126,000	378,069
USSR	52,000	–	52,000
Yugoslavia	5,514	–	5,514
Zimbabwe	3,750	4,898	8,648

Annex 6 (Contd.)

Bilateral Donors	1984/85	1986	Total
UN Organizations			
FAO	–	–	–
UNDRO	–	–	–
UNHCR	7,600	–	7,600
UNICEF	6,760	–	6,760
WHO	–	–	–
WFP	13,994	21,856	35,850
Private Charity			
Total from Private Foreign Charity	–	22,000	22,000
Grand total	594,124	290,300	884,424

Note: Total non-food aid 1984/85–US$401,119,000.
Total non-food aid 1986 –US$188,790,000.
Grand total of food and non-food aid
1984/85–US$995,243,000.
1986 –US$479,090,000.
(The dollar value of both aid and non-food aid is given by the donors themselves, hence the total aid is inflated to some extent).

Source: UN Office for Emergency Operation in Ethiopia; UN Emergency Prevention and Preparedness Group in Ethiopia.